The Male Lumpectomy

Focal Treatment for Prostate Cancer

by

Gary Onik, M.D. and Karen Barrie, M.S.

authorHOUSE®

AuthorHouse™
1663 Liberty Drive, Suite 200
Bloomington, IN 47403
www.authorhouse.com
Phone: 1-800-839-8640

First published by AuthorHouse 7/23/2008

ISBN: 978-1-4184-9771-2 (e)
ISBN: 978-1-4184-9769-9 (sc)
ISBN: 978-1-4184-9770-5 (hc)

Printed in the United States of America
Bloomington, Indiana

This book is printed on acid-free paper.

This book is not medical advice, nor intended to substitute for such. It is not the author's intent to give medical advice contrary to that of an attending physician. For questions of a personal medical nature, please consult your physician.

To the courageous women (including my lovely sister-in-law Anita) who in the face of their own mortality and tremendous resistance from the medical community, took the risks necessary to validate the concept of a lumpectomy for breast cancer.

Table of Contents

Forward

There is a long-standing conviction among physicians that when treating potentially curative prostate cancer the entire gland must be treated, because it was discovered long ago that prostate cancer is "multifocal." This assessment meant that the gland usually contained more serious cancers than were evidenced by the biopsy that confirmed the diagnosis of cancer, and such was certainly true before the PSA test became available. In fact, it is worthwhile to bear in mind that over half of all men diagnosed with prostate cancer before PSA testing became widely used, presented with advanced disease and no chance of cure.

PSA as a screening test has changed all that. We can now find prostate cancer much earlier, i.e. small, low-risk cancers that may be of no threat to life and well-being. Although some of these cancers may still be "multifocal," the question arises whether treating the whole gland with surgery, radiation, or cryotherapy – thereby risking unpalatable and crippling complications – is really necessary. You will find informed and measured potential answers to this question in this very well written book by Gary Onik, *The Male Lumpectomy: Focal Therapy for Prostate Cancer.*

The book, one of its kind, is informative, masterfully composed, thought-provoking, and reads easily. It contains a wealth of wise, unpretentious advice, deftly and agreeably written, and his conclusions are increasingly supported by empirical evidence. Dr. Onik is able to relate and explain difficult subjects as if he is face to face with the reader. When he uses medical terms he refers to examples rather than technical explanations. In the often complicated world of medicine, we should learn to listen more often to the patient's voice, which Dr. Onik superbly illustrates by using enriching quotes from patients. His courage to go public with his deep-founded doubts about conventional management of prostate cancer, which advocates radical intervention for even the smallest cancerous lesion, brings to mind one of Robert Frost's poems, "The Road Less Traveled," the last stanza of which reads:

I shall be telling this with a sigh
Somewhere ages and ages hence:
Two roads diverged in the wood, and I--
I took the one less traveled by,
And that has made all the difference ...

Dr Onik's commitment to revise long-held assumptions and beliefs promises a new and fresh approach to prostate cancer along a "Road Less Traveled," which I believe will have an increasing role in managing this disease. I can recommend the book enthusiastically and without reservation.

Haakon Ragde, MD
Executive Director
The Haakon Ragde Foundation for Advanced Cancer Studies
Clinical Professor of Urology
University of Virginia Medical School

Introduction

This book is about a coming revolution in prostate cancer treatment, a "Male Lumpectomy." For the first time in history, medicine can offer a targeted approach to treating just the cancer, instead of destructive whole gland treatments that potentially create havoc. In fact, the conventional "radical" approaches leave men impotent and/or unable to maintain control of urination more often than physicians admit.

This book is about a solution to the dilemma that many prostate cancer patients have been facing. I expect these concepts to be very controversial within the urologic community and I accept this as natural reaction to a major change in medical thinking. In my career as an interventional radiologist I have seen many instances of groundbreaking new therapeutic approaches meet significant resistance (often made more heated by turf battles among medical specialties) only to be eventually accepted as the standard of care. A technique called Transluminal Angioplasty is one such example, and even now many gynecologists are accepting a new minimally invasive, radiology-developed technique for treating uterine fibroid tumors called Uterine Artery Embolization (UAE). As patriot Thomas Paine penned in his pamphlet *Common Sense*, "Time makes more converts than reason."

This idea of a targeted prostate cancer treatment did not jump out of thin air. It is very similar to the progression of breast cancer treatment from radical mastectomy to the much less invasive breast lumpectomy. Treating breast cancer with radical mastectomy surgery was based on the faulty assumption that cancer metastasizes (spreads) in an orderly manner down a logical path from cancer mass to lymphatics, through the tissue to lymph nodes, then finally gaining access to the blood stream where it spreads to other organs. Thanks to new clinical evidence, the rejection of this "train track" breast cancer spread made radical mastectomy yield to the concept of the breast-sparing lumpectomy. No longer was it necessary to remove breast, muscles and lymph nodes to cure a patient with breast cancer. Though slow to happen, embracing this alternative treatment was driven by women's desires to preserve their breasts. Today, lumpectomy is the predominant treatment for breast cancer, and its effectiveness equals that of complete breast removal.

It is ironic that the idea of a lumpectomy for prostate cancer has lagged almost 25 years behind this breast cancer breakthrough, since urologists already accept the notion that the survival of prostate cancer patients is determined by the biology of their disease rather than the local cancer therapy they receive. Radical prostatectomy persisted in the face of this change in thinking because to remove only a portion of a prostate gland by surgery was not technically feasible. Additionally, there was a mistaken belief that prostate cancer involved *all* portions of a gland in *all* patients. It wasn't until image guided cryoablation (the killing of a tumor by freezing it while still in the body—dead tissue is then removed by the body's natural defense system) proved its utility in long term studies, that the ability to treat just a malignant portion of a prostate gland became a practical reality. This then prompted a closer look at the pathology literature, which revealed that upwards of 75-80% of prostate cancers might be amenable to a lumpectomy type approach to prostate cancer. The stage was finally set for "The Male Lumpectomy."

I do not take my responsibility in advocating this new treatment approach lightly. When I treated the first prostate cancer patients with a lumpectomy, I knew that both the patients and I were taking a calculated gamble. Understandably and quite correctly, one could say that it was the patient who was taking the real risk in this bargain, and that it was easy for me to perform an "experimental procedure" on patients. This is false. I identify with my patients and try at all times to practice the golden rule of medicine, "treat patients the way you would want to be treated." When I treated the first men with this new procedure, I believe I can say with honesty that had I been faced with the same choice as a patient, I would have proceeded along the same course. In fact at one point in my career I was faced with a very similar situation in which I had to decide whether I would undergo one of my own experimental procedures. At the time I was a Professor of Neurosurgery at the Medical College of Pennsylvania and I was developing a new procedure to treat herniated lumbar (slipped) discs. The procedure involved a laparoscopic approach to the spine from the abdomen in which the herniated disc is removed from the front. We had treated five patients successfully when I developed a herniated disc myself. Faced with the choice I did decide to have my own experimental procedure performed by one of my colleagues (unfortunately it was unsuccessful and I went on to have the usual more invasive spine operation). I therefore feel that I have earned some credibility in this regard.

I also understand that beyond the individuals I treat, the lives of thousands of other men could be at risk if I am wrong. This is one of the main reasons I waited to see long-term results in my own series before publishing the first articles in the medical literature, thereby opening the concept to investigation by other researchers. Also coming into play was the fact that only now was the technology to perform a reproducible male lumpectomy available for widespread use and evaluation. Reproducibility is the key to any medical advance and I can assure you that the nightmare of a medical researcher is to find that others can't reproduce his data. Only now has the procedure of cryoablation become standardized enough to leave the realm of an art and enter a consistent technology-driven science. This is not to suggest that the Male Lumpectomy is a finished product with all the data available to make it the preferred choice for all patients. I believe, however, that we are to the point where patients wanting such an approach are taking a reasonable risk in entering the studies that are now being carried out on this treatment approach. It is the courageous breast cancer patients who forged the way for men to take this measured risk in striving for a marked improvement in the care of prostate cancer.

The purpose of this book is to explain the logic, medicine, science and compassion behind a tailored approach to prostate cancer treatment. It is not meant to hype a new fad or trend like a "revolutionary new diet book." It was written to inform prostate cancer patients that the concepts about prostate cancer are changing and there is a wider range of options. Only they can decide if they feel comfortable being one of the early patients who choose this option, becoming part of medical history in the process.

Gary Onik MD

Chapter 1
A New Approach To An Old Problem

Why wasn't this around when I needed it?
Anonymous Prostate Cancer Survivor

"I'm sorry to say this, but you have prostate cancer."

As the initial shock wears off, today's newly diagnosed man is likely to mobilize his own search for his own best treatment. No doubt his doctor has made a recommendation, but he wants to know what else is out there. He likely boots up his computer and follows links from one prostate cancer website to another. He may attend a prostate cancer support group where he hears others' experiences and picks up materials. He learns that surgery, radiation, cryotherapy (also called cryoablation, cryosurgery or simply cryo) and watchful waiting are his four main choices, with variations of each. He learns that all treatments carry a risk of temporary or lifetime side effects, some humiliating, even if they eradicate the cancer. He learns there is no guaranteed cure, and if his treatment fails he may have few "salvage" options. He logically conducts a cost-benefits analysis, while he instinctively fears that his body and even his life will be mutilated in some way. He ponders how much time he has before he must decide, and hopes that tomorrow will bring news that the cure for cancer has been found.

A new vision of prostate cancer treatment

The magic bullet hasn't arrived, but a groundbreaking treatment has! I call it the **Male Lumpectomy** or cryolumpectomy. The term *cryo* comes from the ancient Greek word *kruos* that means cold. The Male Lumpectomy kills the "lump" or tumor by lethal frostbite yet spares the portion of the prostate gland that is not frozen. No other treatment besides the Male Lumpectomy can conform cancer-killing power precisely to the tumor and surrounding at-risk tissue while preserving the healthy tissue and one or both nerve bundles that control erectile function. As you will see, this promising revolution in prostate cancer treatment is **minimal yet radical**.

This book is based on concepts that were forged while establishing the lumpectomy for breast cancer, my research, input from other professionals, and the success stories of my patients. Since investigation of the "female lumpectomy" for breast cancer had a head start of 30 years, applying the same principals to prostate cancer treatment has a bit of catching up to do. Clearly, the Male Lumpectomy is **a new vision of focal prostate cancer treatment** whose time is here.

From mastectomy to breast lumpectomy

Cryolumpectomy is great news for men, like a similar breakthrough in women's health care. The story starts with a **dramatic revolution** in breast cancer treatment.

The similarities between breast and prostate cancer are striking. Each year there are over 200,000 new cases of both. Long-term survival is improving, but each claims more than 30,000 lives annually. In some families, the two cancers are genetically linked. Both involve sexuality and identity. Men fear loss of masculinity just as women fear loss of femininity. But unlike some prostate cancer patients, women with breast cancer can't "watch and wait." Their time bomb is too deadly.

Breast cancer treatment affects a woman's identity. Her breasts are an essential part of her body image, sexuality, maternity. It was long held that the best way to prevent cancer spread throughout the body was **radical mastectomy**, or the removal of the entire breast along with muscle and lymph nodes into the armpit. A double mastectomy is traumatic, leaving a woman with no breasts and often impaired arm mobility and strength. There are different cell lines of breast cancer, just as with prostate cancer, but radical breast surgery does not distinguish among them. A chance at life in exchange for breasts—a bitter tradeoff, yet somehow survivors adapted to flat, scarred chests. Women feared radical mastectomy, but what choice was there? Doctors were saddened by the mutilating consequences of mastectomy, but what choice was there?

The clinical value of radical mastectomy was based on the belief that breast cancer spread as if traveling down a railroad track from breast tissue to the lymph nodes, then on to distant sites. The dawn of a new understanding about the biology of breast cancer brightened the treatment landscape. Research pioneered by surgical oncologist Dr. Bernard Fisher almost three decades ago showed that the traditional idea was mistaken. Large cooperative randomized studies (the "holy grail" of cancer research) definitively proved that the degree of local treatment made no difference in life expectancy. Whether patients had cancerous lymph nodes removed or not (the difference between radical and simple mastectomy) had no effect on their life or death. Apparently the cancer was not spreading down this train track in an orderly trip to another organ, but was instead taking a direct plane flight based on biological factors not fully understood. It was clear, though, that stopping the train at an early station by a radical, mutilating procedure was not going to affect survival. Thus, the way was opened to the **lumpectomy concept** of local breast cancer control. As Dr. Fisher summed it up, "After 20 years of follow-up, we found no significant difference in overall survival among women who underwent mastectomy and those who underwent lumpectomy with or without postoperative breast radiation."[1]

The first brave doctors and patients who were willing to risk removing just the lump and its margins had an uphill struggle in the face of resistance and criticism. Women were told, "Radical mastectomy is the gold standard." That sounds very authoritative—what does it mean? It means it is the approach with the longest data, so it sets the bar for measuring the success of new approaches. It doesn't mean it's the best or only option, since subsequent advances eventually catch up with or even exceed the shiniest "gold standard."

Today **lumpectomy**, not radical mastectomy, is the most common form of breast cancer surgery. Depending on the size and location of the tumor, as much breast is spared as possible. Radiation is then used as a **safety net**. The patient is carefully monitored to make sure cancer has not returned, and treated early if it does. Lumpectomy followed by radiation is **as good as mastectomy** when the tumor is **unifocal** (found in only one site) and is less than four centimeters with clear margins (no cancer in surrounding tissue.)

While lumpectomy is only appropriate for breast cancer patients who meet certain conditions, if cancer is **detected**

[1] Fisher, Bernard et al.. "Twenty-Year Follow-up of a Randomized Trial Comparing Total Mastectomy, Lumpectomy, and Lumpectomy plus Irradiation for the Treatment of Invasive Breast Cancer" *New England Journal of Medicine* 16:347 (Oct. 17 2002) 1233-41.

early those conditions are more likely to exist. Women are advised to explore all options before making a decision:

> Doctors in some parts of the United States may be more old-fashioned and less likely to offer lumpectomy with radiation as an option for their patients, particularly their older patients. Such doctors may urge mastectomy, even for women who should be offered the choice. Don't let hidden biases or unchanging attitudes keep you from getting the best care. Find a surgeon who keeps current, and whose techniques aren't limited to what used to be the standard of care twenty or thirty years ago.[2]

What influenced breast surgeons to change? There were two factors: the willingness of patients to undergo a treatment based on early evidence, and the affirmation of its effectiveness by sound research as the number of cases grew.

This raises two key questions: Can prostate cancer be successfully treated by applying the principles of breast lumpectomy? If so, what advantages would it offer over current treatments?

---

[2] www.breastcancer.org

So far you have seen that:
- **Today's patients** are more likely to conduct their own search for treatment choices.
- Finally men have a breakthrough treatment option called the **Male Lumpectomy**.
- The precedent for it was set by **changes in breast cancer treatment**.
- **Patient demand and growing research evidence** brought about those changes.
- Breast lumpectomy is as good as mastectomy for women with **unifocal tumors**.
- Male Lumpectomy may **offer men hope** as breast lumpectomy did for women.

Total prostate destruction has a downside

Just as with breast cancer, the conventional approach to killing prostate cancer is total eradication. When a woman loses a breast, the impact on her self-image is the result of an external and visible loss. When a man loses a prostate gland to surgery, radiation or ice, there is no dramatic visible loss. The impact on his self-image is the result of possible damage to urinary, rectal and sexual function. Breast cancer treatment can damage a woman's appearance or ability to nurse a baby; prostate cancer treatment can damage a man's ability to maintain his personal hygiene or enjoy sex. Thus, self-esteem is at risk.

To understand the challenges in treating prostate cancer, let's take a look at **male pelvic anatomy**.

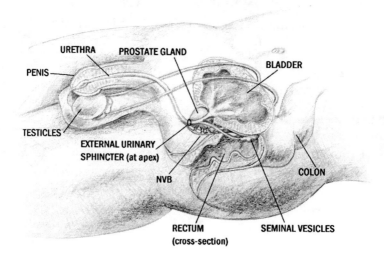

1. The **prostate gland** is a walnut-sized gland nestled below the **bladder** and in front of the **rectum**. It produces fluid that helps sperm survive in the female genital tract.

2. The **urethra** carries urine from the bladder through the penis and out of the body. It passes through the prostate gland. Because it surrounds the urethra, the prostate has some muscle fibers that squeeze it slightly and help control urine flow.

3. The **external urinary sphincter**, or small muscle that controls voluntary urine flow, sits at the gland apex where the urethra exits the gland and enters the penis.

4. The **seminal vesicles** lie above and behind the prostate gland and are considered extensions of it. These glands produce most of the volume of the semen, which mingles with the fluid produced by the prostate at the time of ejaculation.

5. The **neurovascular bundles (NVB)** are connected to the prostate. They hug both sides of the gland like two fine webs. These nerves control **erection** but **not orgasm,** which is an independent function of nerves in the spinal cord.

> **So far you have seen that:**
> - Treatments to cure prostate cancer have been aimed at the **entire prostate gland**.
> - The prostate touches both the **external sphincter** (responsible for urinary control) and the **neurovascular bundles** (required to have an erection).
> - Side effects can have **physical, psychological and lifestyle consequences**.

Deferring treatment: buying time or fear of commitment?

We don't know exactly how many different prostate cancer cell lines exist, but we know they don't all behave the same way or respond equally well to all treatments. Which type of treatment works best with each cell line? No one knows.

Prostate cancer used to be called an "old man's disease," a byproduct of the cell breakdown that accompanies aging. Prostate cancer has long been believed to be both **multifocal** (cancer scattered throughout the gland) and **slow-growing**. While these characteristics are still true in many cases, new findings are challenging both assumptions and their implications for managing the disease. Younger men in their 40's and 50's are now being diagnosed. African-American men are at greater risk of developing prostate cancer than other ethnic populations; for reasons that are not clear, they are also more likely to be diagnosed at an advanced stage, and are more than twice as likely to die of prostate cancer as white men. Many researchers theorize that some prostate tumors result from exposure to pollution, chemical fertilizers, pesticides and other toxins that may generate **environmentally-driven** prostate cancers, affecting men of all ages. For instance, the U.S. government has acknowledged that many Viet Nam veterans in their 50's have prostate cancer as a result of exposure to the herbicide Agent Orange. These cancers may be **more aggressive**. Also, the male hormone testosterone seems to accelerate cancer growth, and younger men generally have **greater testosterone levels**. Finally, younger men with a **family history** of prostate or breast cancer are being diagnosed sooner. Whether they get cancer younger because they're at risk, or find it sooner due to better testing, is arguable. The point is that ethnicity, environment, age and heredity all affect the chance of men developing prostate cancer at different ages and with different degrees of aggression. It is now certain that prostate cancer is not exclusively a slow growing, old man's disease.

It is also important to stage the cancer accurately at the time of diagnosis. Patient support organizations suggest getting a second opinion on biopsy slides because there is a risk of underestimating or **under-staging** the cancer's aggression. If this occurs, a treatment may be chosen that is insufficient. Relying on outdated assumptions or inaccurate diagnosis can lead to the illusion of security, even denial.

Let's face it, a lot of guys avoid going to the doctor if at all possible—it's a Guy Thing. If they have an excuse to put off having a procedure, especially one that might put masculinity at risk, they grab it. It's important to be rational in the face of fear, but it's tempting to rationalize instead. Believing that the cure is worse than the disease may have a high price tag, especially because prostate cancer initially announces itself with just a quiet little "whisper," a rising PSA level (prostate specific antigen) assessed by simple blood tests. If men wait to seek help until prostate cancer "speaks up" through symptoms—difficult urination, pelvic pain, or bone pain—they have waited too long. It's almost certainly too late for cure. Since most men don't have any symptoms when diagnosed early, holding off on treatment seems reasonable. However, one must keep in mind that although this cancer may be slow growing, it is still growing.

Medical professionals and public health officials debate when it's best to test for and treat prostate cancer. Patient support organizations like *Us Too! Prostate Cancer Education and Support* promote early detection, often called screening. The earlier prostate cancer is diagnosed, the more treatment options and better success odds a patient has. On the other side, government watchdog agencies and medical associations are concerned about rushing men into expensive tests and treatments. They point out the cost to taxpayers, and argue that there are more post-treatment side effects than doctors admit.

A concerned layman, Terry Herbert, has a website where he articulates the problem. He quotes Stanford cancer expert Dr. Thomas A. Stamey:

> As long ago as 1994 Dr. Stamey was voicing his concerns… In a summary of the position as he saw it then, he said: 'To the extent that perhaps as many as 50% of these procedures are either unnecessary (the cancer is too small) or ineffective (surgical excision of the prostate fails to cure the patient of his cancer), they represent an enormous cost to an already overburdened health care system…' Stamey was therefore suggesting

that as many as 95,000 men - around 8,000 men every month - would receive treatment that year which was unnecessary or ineffective. All would suffer side effects (technically called morbidity) from this treatment. For every man in that position there would likely be… family members, all suffering to a greater or lesser extent. And all for nothing, if Stamey is correct.[3]

To treat or to wait? If the male equivalent of a lumpectomy can defeat the tumor without the side effects of whole gland treatment, it would be a true happy medium, resolving the dilemma of an all-or-nothing approach!

So far you have seen that:
- There are **different lines** of prostate cancer cells.
- Age factors, environmental toxins and genetic links may result in **high-risk cancer.**
- Doctors and patients **may not know** which cell line a patient has.
- Some high-risk tumors are **under-staged** and **undertreated,** causing treatment failure.
- Some low-risk tumors are **overtreated,** needlessly endangering quality of life.
- Many men **put off treatment** because they fear "the cure is worse than the disease."
- Men need a **happy medium** that kills the tumor without harming quality of life.

Men's Voices

My follow-up appointment was a shocker. I was an active 59 and too young, in my mind, to have an 'older guy's' disease. This kind of cancer was a bit

---

[3] (http://users.kingsley.co.za/~ghanesh/)

more aggressive than some. It wouldn't wait around for me to age. All at once I was learning about Gleasons, staging, and techniques for ridding myself of this interruption. And that's how I saw it...It was a pest - an intruder - and I had no intention of allowing it to make my life miserable. (DS, California)

After dealing with the trauma of the news I began an intense study to understand the options for treating the disease. (LJ, Florida)

My pulmonologist had recently advised me against having any abdominal surgery and general anesthetic except in a life-threatening situation. So when Doctor G. wanted to go in 'next week' I hesitated. At my hesitation, Doctor G. said, 'You're too young for radiation, so we have to operate.' Still I didn't jump at the chance for surgery, so after a bit of wait, he said, 'Well, you could just wait a while and watch it to see how it goes.' I asked him how long it might be before I had to do something and he replied, 'Oh, five years, maybe ten.' That seemed a bit different than having it cut out 'next week.' Then I asked how I should go about monitoring the disease and he said, 'Get a PSA every year, or maybe even every six months and if the PSA doubles in a year or gets to 20, get it taken care of.' I thanked the doctor and told him that I would get back to him. On the way home, my wife and I discussed what we had just heard the doctor say, and by the time we got home we had agreed that I would opt for what he called watchful waiting. I would get a PSA test every 3 months, would not allow it to go above 15, and would do "something" within two years. At that point I began to feel as though I were in control, and if subsequent events proved that I had made a bad decision, it was my decision, and I could blame no one else. Deep down, I was hoping for a miracle cure of some kind. (DM, Michigan)

Summing It Up

Men feel about prostate cancer the way women do about breast cancer. When they are diagnosed, they fear that life as they know it will depreciate as a result of treatment. For men, their urologists, radiologists and cryosurgeons can seem like well-intentioned car repairmen who claim they can fix the tick in the Ferrari's timing chain but they might have to bang up that sleek front end or sacrifice the hydraulic line in the process. Many owners drive off, hoping the chain won't blow before they get where they're going. The cure seems worse than the disease.

Hope is just around the corner. There's a new group of mechanics on the block. We* have a promising solution that can fix the tick without gumming up the works. Using the principles learned in treating breast cancer we have pioneered a form of cryotherapy that freezes less than 100% of the gland to avoid denting a man's identity. Just as surgical lumpectomy was a new vision of breast cancer treatment, tailoring a freeze to a tumor is a new vision of prostate cancer treatment.

To get a full grasp of why a happy medium is needed, the next chapter examines the possible side effects of total gland eradication and the specific impact of each treatment.

---

*Dr. Onik's Note: Throughout this text I often interchange the pronouns "I" and "we" when explaining the evolution and scientific application of prostate cryotherapy. I am proud to be a contributor to this exciting treatment development, yet I am not alone. Despite occasional media attention of a celebrity doctor, no major medical advance can rightfully be attributed to one person. Patients courageous enough to risk new treatments, engineers who develop new equipment (particularly Boris Rubinsky, Ph.D., my long term collaborator) and other physicians who collaborate to test and refine the treatment concept are all integral to the process. When I say "we" it is to all these fellow contributors, known and unknown, that I gratefully refer.

Chapter 2
The Wish For A Better Solution

As to diseases make a habit of two things—to help, or at least, to do no harm.
Hippocrates

All of us like to believe that when stress occurs we are rational. This means being able to calm our natural fight-or-flight reactions in order to gather data, define the problem, logically evaluate action alternatives, make a balanced decision, and accept the consequences. The greater the stress, the greater the challenge to rational thought.

A diagnosis of prostate cancer is not a death sentence, but it does create distress. Some men say they aren't worried, but gallows humor often suggests otherwise. Perhaps you've heard surgeons, radiologists and cryosurgeons referred to as "butcher, baker, ice cream maker." The words *cutting, burning* and *freezing* create anxiety. Knowing that there are options should be reassuring, but the thought of treatment can also cause stress.

As Chapter One pointed out, all approaches that completely eradicate the prostate gland might impair bladder, bowel and sexual function temporarily—even permanently. **None can guarantee cure.** The side effects of prostate cancer treatment are called **morbidities**, a scary-sounding word that means unintended collateral damage. The two most common morbidities have to do with what guys know as "getting it up" and "getting up to go to the john."

Impotence: A Possible Side Effect Of Treatment

The prostate gland is not essential for your own life, but it helps transmit life by manufacturing fluid to help sperm reach their goal. Because it has this vital role in procreation, Mother Nature protected it well. As illustrated in Chapter 1, she tucked it away where treating it can affect the nerve bundles that control potency.

Potency means the ability to have a spontaneous erection firm enough for intercourse (penetration) to both partners' satisfaction. It means "getting it up" without assistance from medication or a mechanical device. **Impotence** means you cannot spontaneously achieve erection sufficient for penetration, or you get a partial but not full erection. A doctor can clinically measure the degree of erection, but only a patient knows if he and his partner are satisfied.

Many men and women confuse potency with **orgasm** or "coming." You may not realize it, but a man can **still have an orgasm** even with a soft penis. *Orgasm* is controlled by a different nerve system located in the spinal cord and is not physically affected by prostate cancer treatment. Many patients experience despair and depression because they don't know this. For some this seems like a poor consolation prize because their pride as lovers is very much tied to their erections. Even when women assure their partners that they can be satisfied in other ways, many impotent men identify sex with erection and are troubled by guilt or embarrassment. It takes a positive attitude and a supportive partner to experiment with pills, pumps or needles. It's not surprising that a patient who believes his sex life will be over may avoid or delay prostate cancer treatment.

Incontinence: A Possible Side Effect Of Treatment

Impotence is not welcome news, but support programs like *Us Too!* assure men that they and their partners can adapt their sexual practice to regain love, intimacy, fun and magic. On the other hand, incontinence is a problem that can ruin a man's life. A man who wears pads or diapers is often afraid to leave the house resulting in a challenging adjustment.

Webster's Dictionary defines continence as the ability to voluntarily control urinary and fecal discharge. **Incontinence** means that a man (or woman, since this problem affects both genders for many reasons) may involuntarily leak urine or feces. Severity is measured by how much protection is needed and how often. Mild incontinence is often called "stress incontinence" meaning that a sneeze, cough or physical activity triggers leakage requiring the precaution of a pad inside underwear. More severe incontinence, however, requires up to several pads a day, or even adult diapers. The same is true for bowel incontinence (most common from radiation, but prostatectomy and cryotherapy can also injure the bowel). It requires acceptance, adaptability and a sense of humor to face temporary or permanent dependence on diapers. While many men and women will eventually have to deal with incontinence as a byproduct of aging, no one wants to think it will happen to them. When a younger man has to face this risk as a result of prostate cancer treatment, it's a sad prospect indeed.

Us Too! informs us that clinical continence criteria vary. It is confusing. Even medical journal articles don't use consistent standards. Continence may be defined as anything from no pads per day to up to four pads. *Us Too!* defines continence as **zero pads** per day. When patients are told that they won't be incontinent after treatment, *Us Too!* urges men to ask their surgeons or radiologists, "How do you define incontinence?"

So far you have seen that:
- **Morbidity (or morbidities)** is the term for side effects of treatment.
- **Impotence** is a possible morbidity, but it does not affect orgasm.
- **Incontinence** is a possible morbidity that affects lifestyle more than impotence does.

The side effects of each local whole-gland treatment

Radical Prostatectomy Surgery, or simply **RP**, is the **surgical removal** of the entire gland, seminal vesicles and pelvic lymph nodes. *Note: RP cannot be a "partial" approach to just part of the gland.* It may be done through an incision in the abdomen or perineum (skin between the scrotum and anus), or by abdominal laparoscopy with a camera and instruments inserted through several small incisions. The following illustration shows how removing the prostate affects pelvic anatomy. Note that in this case, removing the gland, seen in the upper left of the illustration, also involved the threadlike neurovascular bundles. Even when one or both bundles are left in the body, the trauma may result in temporary or permanent impotence, as explained later in this section.

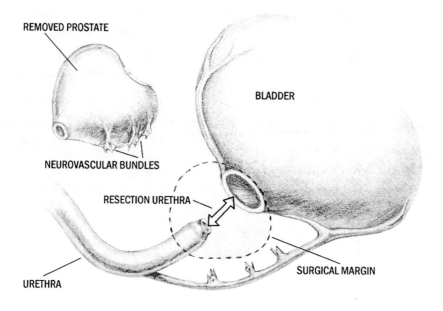

REMOVED PROSTATE

NEUROVASCULAR BUNDLES

BLADDER

RESECTION URETHRA

URETHRA

SURGICAL MARGIN

Surgical removal (RP) requires reconnecting the urethra.

Cutting out the gland removes the part of the urethra that passes through it, requiring delicate handling of the urinary sphincter muscles and reattaching the ends of the urethra to each other. This is called resectioning the urethra. There is risk of urinary incontinence due to trauma to these structures. Also, some patients report reduced penis length, possibly due to this. Because of the blood vessel complex in this area there is a risk of blood loss. Patients often bank a couple pints of their own blood before RP, which has the longest average recovery time of all options.

Depending on tumor aggression, the surgeon also cuts away part of the bed of tissue around the gland to prevent the spread of cancer that may have penetrated the gland margin. During the surgery, he/she must also determine if the nerve bundles that control erection can truly be spared. It is difficult for prostate cancer to penetrate the sheaths around the nerves, but if it does it can spread swiftly to other organs along this "superhighway". The surgeon may find that he has to remove one or both nerve bundles after all. Even if he decides it's safe to spare them (recent studies have shown that surgeons can be mistaken in this determination) this delicate work may damage them. After nerve-sparing RP, some men are potent right away, while others find that potency returns slowly or not at all, despite assurance that the nerves were spared. **Remember**: RP does not offer the option of partial gland removal. The <u>whole gland</u> comes out. A man considering nerve-sparing RP should attend support groups beforehand, or sign up for online Prostate Cancer Mailing Lists [see Appendix I] where he can ask men who had RP what their sex lives are like. Most prostate cancer survivors are honest about post-treatment sex. Impotence following RP may be treated with artificial erection assistance. Orgasm with or without erection is possible as a function of spinal cord nerves, but will be "dry" with no prostatic fluid.

The latest innovation in RP is the laparoscopic or "robotic" radical prostatectomy. This minimally invasive approach to RP minimizes the incision needed for the operation and certainly seems to improve the recovery time associated with it. Still, robotic RP has all the risks associated with open RP such as impotence, incontinence, and positive surgical margins (cancer left behind). Whether or not robotic RP will lower the incidence of these complications is yet to be determined, but one thing is clear: it has not eliminated them.

Surgery appeals to patients who strongly feel, "I want the cancer OUT!" The success statistics of RP in removing all the cancer, however, range from 70-92%. Nerve-sparing success ranges from 50-85%. If RP fails (the cancer most likely had already left the gland), external beam radiation to the prostate bed may be indicated if the spread is local. If this also fails, or the cancer has spread beyond the pelvic region, the next step will be hormones. Most surgeons have an age cut-off and won't do RP on older men.

Radiation therapy can be **external beam radiation therapy (EBRT),** including newer options like Intensity Modulated Radiation Therapy (IMRT), proton or conformal beam; or **internal** radiation like **permanent or temporary radioactive seeds** (brachytherapy, brachy, seeds.)

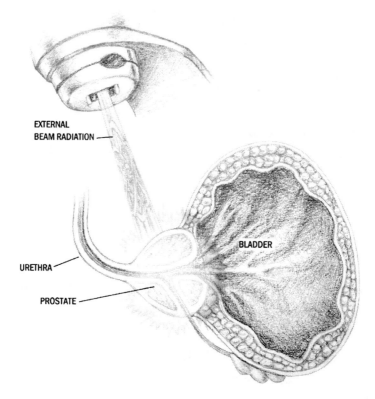

EXTERNAL
BEAM RADIATION

BLADDER

URETHRA

PROSTATE

Beam radiation, IMRT, proton beam, etc. radiates the prostate as accurately as possible though there is always some scatter.

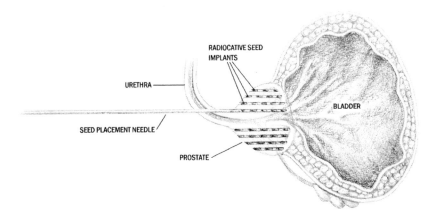

Radioactive seeds (brachytherapy) are implanted throughout the gland; permanent seeds may affect nearby structures over time due to radiation scatter.

Radiation does *not* destroy cancer cells right away, so exposure over time is necessary. That's why EBRT can take up to seven weeks, and why seeds are left in place inside the gland. Radiation works by damaging cancer DNA and interfering with its ability to reproduce, but if some cells survive they can mutate and become more aggressive. Generally, the higher the cancer risk level, the more radiation needed. Radiologists must balance the effective killing dosage against the risk of tissue damage to other structures. Newer radiation technologies have greater accuracy of focus, but all radiotherapy has some scatter effect so nearby structures are at risk of gradual damage. Some radiation side effects may show up soon after therapy, others later.

Radiation is minimally invasive, but during treatment it is not possible to monitor its accuracy and success. Cancer cells die gradually as they become unable to reproduce. Brachytherapy can also temporarily affect other people. After permanent brachy, patients are advised to avoid close contact with children and pregnant women for a period of weeks. Some brachy patients temporarily wear lead-lined underwear to bed with their wives or partners. Patients are initially continent and potent, but permanent urinary, erectile or rectal problems may develop over time after radiation therapy. There is no way to protect the urethra or external sphincter from radiation. If damage develops, it's often a urologist, not the radiologist, who deals with it. If impotence occurs, patients can still have orgasm even though they will not have an erection, or need assistance with erection.

If radiation fails and cancer comes back (recurrence), it can be more difficult to treat than recurrence following surgery or freezing. The rule of thumb has been that no more radiation can be given, though some centers now offer salvage radiation. However, options are limited. RP is usually not advisable, since the radiation has "fused" much of the prostate bed making removal risky, though not impossible. Salvage cryotherapy is an option if the cancer has not escaped beyond the prostate bed, but it is associated with a greater chance of complications than if the cryosurgery was accomplished without the radiation previously given. As with RP failure, if the cancer has left the region, the patient will be placed on hormones. Remember: cancers that survive radiation may have mutated into more aggressive forms, so salvage treatment should be quickly sought if there are three successive rises in post-radiation PSA.

Radiation appeals to those for whom sexual function is important, at least initially, and men who don't want/can't have surgery. It has a drawback: there is a maximum lifetime threshold of radiation exposure in the pelvic area, so as much as possible is given the first time. Prostate cancer patients with a known risk for other cancers (colon, etc.) may want to save radiation for a future need. Success for radiation ranges 60-92% depending on the tumor type and risk level (low, moderate or high). Side effects include urinary difficulties, late-onset impotence, and bowel problems. Recurrent cancers may be more aggressive than the original.

Cryotherapy (cryo) of the whole prostate uses minimally invasive freezing to destroy the prostate gland, margins, and adjacent prostate bed. Slender probes are inserted through the skin of the perineum into the gland.

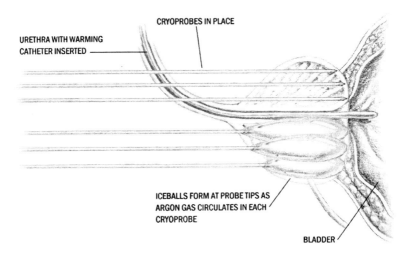

Placement of probes and iceballs beginning to form at tips.

Pressurized argon gas is circulated through the sealed probes, turning the treated area into an extremely cold mass that ruptures and dehydrates cancer cells and the blood supply that feeds them so they cannot recur. The computerized process precisely tailors the iceball slightly larger than the gland so it freezes the nerve bundles and margins as well. It kills any cancer cells that have begun to spread beyond the gland into the prostate bed. In higher risk patients, the doctor also treats the seminal vesicles if the biopsy showed cancer there. Unlike radiation, ice kills all cell lines regardless of aggression. Long-term data shows that whole-gland cryo is as successful as other treatments for localized low risk cancer, more successful for moderate risk, and has the highest success rate with high risk tumors.

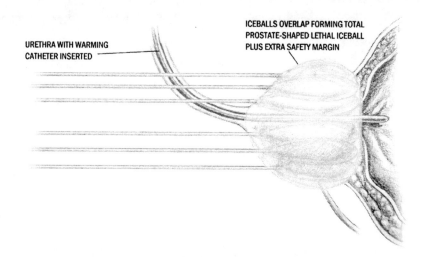

URETHRA WITH WARMING CATHETER INSERTED

ICEBALLS OVERLAP FORMING TOTAL PROSTATE-SHAPED LETHAL ICEBALL PLUS EXTRA SAFETY MARGIN

A completely formed iceball encompasses the total gland.

Since the iceball has precise boundaries, the doctor monitors its location, size and temperature. Cryo differs from RP and radiation in its ability to protect the urethra, which is neither cut nor radiated; it is protected by a warming catheter, resulting in the lowest average incidence of incontinence (less than 2%). Patients recover quickly. However, because the nerve bundles are frozen during total cryo, the ability to have spontaneous erection is halted for up to 95% of patients— the highest average rate of impotence of all treatments—but 47% of men who were potent before cryo return to having sexual intercourse with time. This is the reverse of radiation, in which 50% of men experience decreasing potency over time. As with RP and radiation-induced impotence, erections can be assisted and a dry orgasm is still possible. Some men report reduced penis length, possibly due to the shrunken prostatic scar tissue left after cryo.

If local cancer recurs, cryo is repeatable (salvage cryo); radiation and RP are also salvage options.

Total cryo appeals to patients who want a rapid return to normal activity, who don't want RP or radiation, or who can't have either for medical or age factors. It also appeals to men for whom sexual function is not a strong consideration. Cryo is a top priority treatment for patients with high risk localized disease. "Salvage" cryo is Medicare approved specifically for radiation patients whose cancer has recurred.

So far you have seen that:
- **No treatment** (prostatectomy, radiation or cryotherapy) guarantees a cure.
- All whole gland treatments **carry some risk** to bladder, bowel or sexual function.
- **Surgery** has the highest risk of urinary incontinence.
- **Radiation** has the highest risk of greater aggression if cancer recurs.
- **Total gland cryo** is at least as effective as all other therapies. It has the lowest risk of incontinence but the highest risk of initial impotence.

The choice to leave the cancer in the body

In a recent issue of *PCRInsights*, the newsletter of the Prostate Cancer Research Institute, oncologist Mark Scholz MD wrote:

Local treatment options (radical prostatectomy, brachytherapy, external radiation, cryotherapy) are directed at the eradication of the prostate gland and the cancer it contains. Modern technology in expert hands can accomplish the sterilization of the prostate gland from cancer with a high degree of consistency. However, there are two potential drawbacks... One is the potentially irreversible side effects to the adjoining structures (e.g. the erectile nerves, bladder or rectum). The other is the disheartening possibility of undergoing the risks of local therapy only to later have a relapse because the cancer had already spread to elsewhere in the body...

Side effects of treatment take on added importance, and the quality of life becomes a priority when survival is no longer the central issue. Side effects from treatment tend to be immediate, whereas the slow-growing effects of untreated cancer may not be felt for 10 to 15 years.[4]

In other words, each prostate cancer patient must weigh the **toxicity risks** of the cancer against the **toxicity risks** of treatment. A man who believes his cancer is low-risk and is not ready or willing to seek localized treatment has four options: watchful waiting, conservative management, systemic hormone intervention, or a combination of these.

Watchful waiting (WW) is a calculated risk. The cancer is **left untreated**, but monitored at intervals by the PSA (prostate specific antigen) blood test and DRE (digital rectal exam). Let's take Roger, a 70-year old patient with a PSA of

---

[4] *PCRInsights* 2/2003, Vol. 6 No. 1, pp. 1-6

4.7. His urologist, Dr. Goodbuddy, tells him, "Roger, about that small hard spot on your prostate—your biopsy results came back positive. You have prostate cancer. But don't worry. You're in good health with years ahead of you. Most likely you'll die *with* prostate cancer not *from* it, because it's slow growing. Your PSA is barely above four. We could discuss surgery or radiation, but why? You and Rosie are still sexually active, you're continent—let's not rock that boat prematurely. We'll just check your PSA again next year and see if it's changed. Tell Rosie I said hi." Roger is glad to hear he needn't worry, and goes home to share the news and a little nuptial bliss with Rosie.

In Roger's case, his doctor thinks WW is a good bet because of his age and PSA. It's a practical way for older men or those with low risk cancer to defer whole-gland treatment as long as possible. Each year 30% of new patients, about 70,000, choose WW with or without their doctors' blessing. If PSA begins to rise, or the doctor feels a change during the DRE, another biopsy is done along with diagnostic tests such as bone scan, etc. If the tumor is still biding its time, the man may continue WW. However, as pointed out earlier, a greater number of younger men are being diagnosed—their disease may progress faster. If they choose WW they may miss a crucial treatment window while the cancer is still localized (cure is still possible.)

Correctly choosing WW, however, is very dependent on **knowing the aggressiveness of your cancer and its full extent**. For instance, if your Transrectal Ultrasound (TRUS) biopsy shows you to have a small tumor of Gleason 6 on one side of your gland, you may be a candidate for WW. Evidence is now mounting that both prostate imaging and TRUS biopsy are giving men a false sense of security about the character of their prostate cancer. Studies presented at the 2006 meeting of the Society of Urologic Oncology (Washington, D.C.) by us and other investigators have shown that TRUS biopsy underestimates the Gleason grade in 25% of individuals and the extent of cancer in 50% of cases. We now suggest a new type of biopsy called a 3D Prostate Mapping Biopsy (3D-PMB) for those patients wanting to know the full extent of their disease before deciding between definitive therapy and WW.

Conservative management (CM) is an attempt to boost the body's immune system through lifestyle changes. There is sound research data that healthy diet; nutritional supplements like Vitamin E, selenium, saw palmetto, etc.; regular exercise; and stress management can slow or even prevent rising PSA values. It may go by other names like Active Surveillance, etc. In theory, every person develops random cancer cells but the body's defenses often kill them before they take over. CM is more proactive than WW, hopefully buying time till a safer, less toxic treatment comes along. CM may be a logical strategy for self-disciplined, motivated men of any age with early stage cancer. The problem is that changing daily habits is hard to maintain. "This project is due in two days—overtime again!" "I'll go to the health club tomorrow, I'm too tired now." "What's one more burger this week?" **Cancer is selfish! It seeks its own survival at the price of its host.** If a man can't stick to a healthy lifestyle he's kidding himself, because each cancer cell is like a rebellious teenager with a bad attitude and no impulse control, eager to take advantage if Dad slacks off. There are two downsides to CM: the patient may backslide; or the cell line may not be controllable this way. As with watchful waiting, PSA/DRE surveillance is essential.

Systemic hormonal treatment means introducing hormonal agents (anti-androgens) to suppress or block the action of testosterone. This has an effect on the whole body as well as the cancer, and current thinking is that it is **not curative**. Eventually the malignancy becomes "hormone refractive" which means hormones can no longer stop it. Until then, it keeps the cancer in check by depriving it of hormonal fuel. It may be used for cancer control when localized whole-gland treatment is medically unadvisable, or the patient simply wants to hold off on making that choice. However, blocking androgen production has **side effects**: hot flashes, mood swings, breast tenderness, diminished bone density, and loss of libido. It's like going through female menopause. A prostate cancer patient has to have a powerful desire to trade freedom from potential morbidities for a stage of womanhood even women don't like! Those whose cancer has begun to spread have little choice but to go on hormone blockade. However, new hormone "cocktails" and timing can prolong life for years.

Combining conservative management with systemic treatments is a way to enhance the benefits of each. For instance, intermittent hormone blockade along with nutritional supplements may keep a localized tumor from growing while monitoring it. Men on this regimen whose PSA is stable are motivated to tolerate hormonal side effects. This willingness to accept conditions like hot flashes underscores that patients are often reluctant to go through localized treatment because of associated risks of morbidity.

So far you have seen that:
- The two potential drawbacks of whole-gland treatment are **morbidities** and/or **failure.**
- To hold off on treatment, a man may choose to **live with the cancer in his body.**
- TRUS biopsy and other imaging techniques may **underestimate** the aggressiveness and extent of a patient's disease.

For those with low-risk prostate cancer:
- **Watchful waiting** is a good bet if a man is older and/or vigilant.
- **Conservative management** is a good bet if a man is self-disciplined and vigilant.
- **Hormone treatment** is generally considered non-curative and has side effects that resemble female menopause.

The all-or-nothing dilemma

Is it best to treat prostate cancer at the time of earliest diagnosis? No one knows. Cancer is capable of fast and random mutations. It can "come back angry" after radiation, and "outsmart" hormones. When is the right time to leave it untreated? To eradicate the whole gland? It's all or nothing— and the trade-offs are gambles based on probabilities:

All: Remove or destroy the whole gland in order to be done with the cancer, and risk affecting bladder, bowel and sexual function. Success ranges from 60-95%.

Nothing: Monitor PSA, change lifestyle and/or go on hormones. Local treatment may be a later option, but the tumor may be worse.

So far you have seen that:
ALL = no guarantee, risk impairing quality of life
NOTHING = no guarantee, risk missing a treatment window and shortening life

The wish for a middle ground

When someone comes up with a successful but less toxic prostate cancer treatment, men with this disease will want to know about it just as women with breast cancer did. Physicians are working to improve current therapies. Patrick Walsh, M.D. created a way to spare erectile nerves for qualified RP patients, making it less ominous for men concerned about their sex lives. Laparoscopy made it less invasive and shortened recovery time. 3-D conformal beam radiation and other new delivery systems focus radiation better. Cryotherapy is safer and faster then ever. But **no existing total gland local treatment** offers a satisfying middle ground because the trade-offs are still costly.

Male Lumpectomy resolves the dilemma

Prostate cancer patients need a happy medium that destroys just the cancer without jeopardizing quality of life. A **therapeutic middle ground** must be victorious over prostate cancer and also reduce the chances of potentially negative consequences. We have developed a procedure that will allow thousands of men to have a safe, effective and less toxic treatment. It is the **Male Lumpectomy**, a tailored focal application of cryotherapy. It is the first true middle ground.

> **So far you have seen that:**
> - **No one knows the ideal window** for local gland removal or destruction.
> - A **total gland treatment** that destroys all of the cancer puts **lifestyle** at risk.
> - A **systemic treatment** that destroys none of the cancer puts **lifespan** at risk.
> - An ideal middle ground would meet **two standards**: kill cancer, preserve lifestyle.
> - **Doctors want this** just as much as patients do.
> - We have developed the Male Lumpectomy to offer men a **middle ground**.

Men's Voices

[I attempted] to find a treatment that would not be dangerous to my health, and had a good possibility of either a cure for the disease, or a long remission. (DM, Michigan)

The bottom line is that if I knew what my particular situation was going to turn into, I would not have chosen radical prostatectomy. I followed the statistics forgetting that I was a lousy gambler. Any decision is a gamble, but the next time I hear a white coat refer to the 'gold standard,' I'm going to dump several boxes of printouts on his head. (Anon.)

I was seeded in '99. I still have bowel urgency somewhat frequently so I make sure I'm close to a bathroom or I keep paper with me. (Anon.)

First they put a tube up your penis. Then they punch a hole in your bladder. Next they stick a camera up your behind. Under your scrotum there's a very soft area. They put a little slit in it and insert a thin probe. They line up the probe with the cancer cells, guided by the camera. Then they turn on the gas in the probe to freeze the cancer and that's all there is to cryo. (GN, Scottsdale AZ)

In September it will be six years since my cryo. I've never had a problem with incontinence and sexual activity has returned to a "functional level." My oncologist (Dr. Susan Slimfinger) examined me just last month and proclaimed, "I can't believe it. I think your prostate is smaller than mine." My PSA reading was 0.25. (GG, Fair Haven NJ)

Summing It Up

Here's a tale of a clever man who found a way to satisfy everyone's
   wishes, even his own.

Once there was a poor woodcutter who lived in a tiny hut with his
   scolding wife and her old blind mother. The wife desired wealth,
   and each day sent him to cut wood to sell. But he was a fiddler, and
   instead of cutting trees he hid in the woods and fiddled away.

One day, Old Scratch the Devil appeared and challenged the man to
   a fiddling contest. If the man won, Old Scratch would grant him
   one wish. But if he lost, he owed his soul to the Devil. The man
   agreed, so they appointed the next day to compete.

That night the woodcutter told the two women about the bargain. His
   wife nagged that he should wish for great riches, a big house, and
   lots of jewels for her. His mother-in-law said it would be too selfish
   to do anything but wish her eyesight restored. For himself, the
   man dearly wanted children, for his wife had been barren all these
   years. He lay awake all night, tossing and turning and wondering
   which wish he should choose.

The next day, he met Old Scratch as planned. The devil was practically
   dancing as he anticipated claiming another soul. He played first,
   and he played mightily. The woodcutter was very impressed at
   the smoke that arose from Satan's strings. When it was over, he
   applauded politely. Then the woodsman took up his fiddle. Now
   he played as if his life depended on it—even his soul. All the forest
   creatures came to hear, and the leaves and wind fell silent. When
   he finished, Old Scratch stamped in rage, for the woodsman had
   won.

"Well," said the angry devil, "what's your wish? Better make it good-
   -it's all you get."

The woodcutter smiled, and said, "I wish that my mother-in-law
   might live to see her grandchildren eat off gold plates." His wish
   was granted, and everyone's needs were satisfied.

The next chapter will acquaint you with the history of
curative cancer treatment by ice. The chronology of how
cryosurgery was developed contains the wishes of many, and

a slow progression toward clever solutions. It will also explain how cryo compares with another forward-looking treatment, radio frequency ablation (RF). The tale of how cryoablation evolved in both power and flexibility lays the groundwork for the idea of partial freeze and how it meets many needs.

Chapter 3
This History is Really Cool!

Healing is a matter of time, but it is sometimes also a matter of opportunity.
Hippocrates

For untold eons, humans have known that frostbite is destructive. Most living cells can tolerate near-freezing temperatures for short periods, but too much cold for too long will result in the loss of unprotected body parts through tissue death. Until the mid-19th century, scientists could not generate low enough temperatures to study freezing unwanted tissue for therapeutic purposes, so the history of cryoablation is intertwined with that of physics.

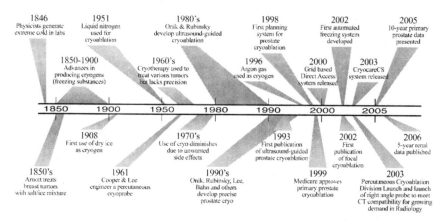

The development of today's cryoablation

Freezing with intent to kill can potentially replace any surgical removal by scalpel. To do so requires a substance called a **cryogen** that can generate lethal cold. You may be familiar with the term "cryogenics," the scientific study of extremely low temperatures. Cryosurgery means applying a cryogen to an organ to destroy it internally instead of surgically taking it out. Before the nineteenth century there was no controlled way to produce cryogens, and no record of anyone using ice therapeutically. Around 1845, English physicist Michael Faraday achieved a low temperature of minus 166° F (-110° C) in his laboratory by mixing solid carbon dioxide and alcohol under vacuum. In the same period, British physician James Arnott published reports of using a solution of crushed ice and sodium chloride (salt) to treat accessible cancers in the breast and uterus. Although his efforts did not completely destroy the tumors, he noted that after thawing, they were less severe and had no discharge or hemorrhage (bleeding). Dr. Arnott is credited as the first physician to attempt cryosurgery. Although his work became part of some textbooks on treating cancer, very few others tried his method.

During the second half of the nineteenth century there were significant advances in producing cryogens, especially ways to liquefy and store gases. These advances spurred a new interest in therapeutic freezing. In 1899 a New York doctor, Campbell White, reported using liquid air to treat various external skin diseases. However, this cryogen was not readily available. Soon after, solid carbon dioxide was found easy to produce and distribute to physicians. In 1907 Chicago dermatologist William Pusey used it, and his success made solid carbon dioxide the most popular method of freezing warts and other skin lesions. It was also used in gynecology, and the term "cryotherapy" was coined. In 1950 a new cryogen, liquid nitrogen, replaced solid carbon dioxide for freezing layers of superficial tissue, but no one had an effective way to treat areas deep in the body, and so it remained until 1960-61 when physician Irvin Cooper and engineer Arnold Lee invented a way to circulate a cryogen in an insulated container with an exposed freezing surface. These men introduced a cryosurgical probe that allowed controlled freezing of inner tissues. It was originally designed for brain applications, but was then recognized as having broader use. Between 1961 and 1970, cryosurgery was successfully used to treat disorders of the brain, uterus, nerves, prostate, and the musculoskeletal system, making it the first truly minimally invasive surgical technique. However, without a good technology to "see" the iceball forming, the effect on the tumor and surrounding tissue was not precise. Because of this, its used decreased during the 1970's as interest in lasers and laparoscopy eclipsed cryo. However, researchers were still curious about cryo.

Destroying tissue with ice inside a living body is complex, based on several factors: the optimal low temperature, how long it is sustained, how rapidly the tissue thaws, and how many freeze/thaw cycles are used. With the correct combination, researchers found, freezing dehydrates and ruptures cancer cells for reliable and effective cell death. They also explored ways to use several cryoprobes at a time to create various sizes and shapes of iceballs. Cryosurgery then became conformal, taking the size and shape of the tumor or gland being treated with remarkable precision! That left the challenge of how to monitor the temperature, and how to "see" the freeze in real time.

In order to measure that temperatures within the iceball were correct, sensitive thermocouples were added to cryoprobes. To monitor safety for body tissue outside the iceball, wire-like thermocouples were developed that could be inserted around the area of treatment. A TV-type monitor provided the practitioner with a graph of temperature changes as they occurred so he could control each cryoprobe accordingly.

A resurgence of cryosurgery occurred in the 1980's. Boris Rubinsky, an engineer and now Professor of Mechanical Engineering at University of California at Berkeley, and I teamed up to tackle the problem of guiding and monitoring cryosurgery. Together we added image-based guidance to observe the ice as it formed. By this time, technology had outstripped decades-old x-ray techniques. Doctors could now peer into the body using an added array of sophisticated tools such as computed tomography (CT), magnetic resonance imaging (MRI), and ultrasound.

Although virtually every known whole body imaging technique can be used to visualize freezing as it happens, Boris and I settled on ultrasound, as an easy-to-use and relatively inexpensive device. Acoustic energy interfaces well with ice because its surface reflects it, and the frozen area looks black in contrast to the softer tissue around it. Here you can see the difference between an unfrozen prostate, and one being frozen:

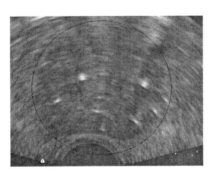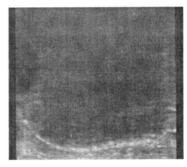

The left ultrasound image shows the cryoprobes (white dots) in place before freezing. The right image shows the solid iceball (black) with the outer edge as a white line.

We ultimately developed systems incorporating thermal monitoring, ultrasound guidance, and the use of multiple cryoprobes to create an iceball that would conform to the individual patient's prostate gland in each unique case. This, along with the invaluable contributions of other pioneers like Fred Lee MD, Duke Bahn MD, Doug Chinn MD and Israel Barken MD became the basis for prostate cryosurgery technology into the new millennium.

So far you have seen that:
- From time immemorial humans have known that **severe cold destroys living tissue.**
- 19th century physicists were the first to **generate extreme cold in the laboratory.**
- **Cryogens** were first used effectively with superficial skin problems.
- **Cryoprobes** invented in the early 1960's enabled freezing deeper in the body.
- **Thermocouples** made it possible to monitor temperature.
- **Body imaging** such as ultrasound allowed doctors to "see" the iceball form.
- See it, get to it, freeze it, monitor it = **cryoablation.**

The future: killing the tumor without removing it

Science fiction is hardly science, but it's not entirely fiction. Look at how many inventions were foretold by novels, movies, and comic books! Have you ever seen a "Star Trek" episode in which diagnosis and healing are accomplished by scanning a patient's body with a hand-held device? Medicine is evolving in that direction. Though early cryo involved major incisions to reveal the tumor site, today's cryoprobes are surgically sharp and very slender. They can be inserted **percutaneously** (through the skin) directly into a tumor.

Small adhesive dressings can replace sutures (stitches) at the entry sites.

Major surgery is far from obsolete, but new ways to destroy tumors are competing with the scalpel. Methods such as High Intensity Focused Ultrasound (HIFU) and Radio Frequency Ablation (RF or RFA) can be used to reduce tumors to clumps of dead cells. As with cryo, they are image-guided. Each has its own advantages and drawbacks.

HIFU is perhaps most like the Star Trek device because the skin does not need to be pierced or penetrated. An ultrasound beam can be focused to destroy tissue very accurately. The major disadvantage of HIFU is that as of this writing, it is not available in the U.S. for the treatment of prostate cancer. However, it is used in some European and Asian countries to ablate prostate and other cancers. Stay tuned!

RF has been used for decades in the U.S. and other nations to ablate cancer by generating an electrical impulse between electrodes at the tip of a slender probe inserted percutaneously (through the skin) into the tumor. Like cryo, RF is being used with prostate, liver, kidney and other cancers. Its advantages include effective tissue destruction by extreme heat ("cooking" as opposed to "freezing"), easy setup, and economical equipment. However, freezing has some distinct advantages over RF.

How ice stacks up against RF

Detroit radiologist Peter Littrup MD has used both RF and cryo for tumor ablation. He points out that the advantages of cryo have to do with the differences between freezing and cooking. The first difference concerns the "architecture" of living tissue. A strong, fibrous protein called **collagen** is abundant in bone, tendons, cartilage and connective tissue. Extreme heat damages collagen beyond repair; freezing does not. Littrup loves practical examples. Take cellophane, for instance. If you hold it over a candle flame, it shrinks, melts, and becomes formless. When you cool it, what's left is unrecognizable, goopy and useless. On the other hand, if you freeze it the basic structure stays intact (but don't bang it when it's frozen or it will shatter). When thawed, it is still cellophane. Littrup recalls his research more than ten years ago with canine prostates. When he examined a dog six months after freezing its prostate he was amazed. "It looked like I hardly did anything." The tissue had regrouped. He found the same thing with lung cryo. "If you freeze a lung tumor a cavity results, but eventually that space fills in."

Using another clever comparison, Littrup describes a brick building with steel girders supporting it. Ablating a tumor with cryo, he says, is like selectively bombing the building. The bricks and mortar are blown away, but the girders remain standing. Masons can come in and fill in the empty space with new bricks and mortar. Thus, for Dr. Littrup, cryo is better for treating lung tumors, since RF can melt nearby bronchial breathing tubes and they will no longer function. Likewise, the trachea and blood vessels can be frozen and regain function, which is not so with RF.

The second plus is how well ice shows up on imaging technology. The picture gives precise feedback. "Oh boy, I'm sculpting this thing *exactly*," is how Dr. Littrup describes his confidence when he's doing cryosurgery. He sees the treatment margin as well as the tumor margin, so he knows precisely where and what he's ablating. On the other hand, RF treatment shows up as tiny tumbling bubbles, because the cooking of tissue releases gases in a sort of boiling effect. Littrup says that during RF, the doctor assumes the tumor is cooked because the extreme heat has generated gas at 100° C., but he doesn't have the same assurance of precise margins that he would have with cryo.

A third advantage of cryo is the shape and size of the lesion it can treat. Cryo radiates out from each probe in a uniform teardrop shape. RF does not have a uniform shape. Ice, with its controllable margins, can safely cover a bigger area because RF can't go beyond a certain temperature curve or charring occurs within the body. Cryo permits a steeper temperature curve, so it goes beyond the cell's ability to protect itself while nearby structures remain intact. These factors—lesion shape, size and temperature gradient—may explain why cryo is more predictable and has lower recurrence rates.

The fourth advantage is the most important from the patient's view. Healing is kinder after cryo than after RF. For cancers such as liver and lung, ice offers much lower complication rates and recovery is much less painful. RF patients often need morphine the first day or two after treatment. Dr. Littrup gives an example with universal meaning: fear of fire. "All creatures have a natural tendency to run away from pain. The residual pain from frostbite is different than from burning. Heat stimulates nerves incredibly well. You see animals run from fire. Even under anesthesia, animals and people twitch during RF. The vital signs [blood pressure, heart rate] change the way they do when people are in pain. During cryo, the vitals don't. RF is initially less costly than cryo because the probes are cheaper and you don't need hookups for pressurized cryogens, but the reimbursement rate to hospitals is high because RF patients must be kept for morphine drip pain management and treatment of peripheral body areas." Patients who have had both RF and cryo for management of bone pain report a vast difference in recovery between the two procedures. Cryo is simply kinder than RF.

Dr. Littrup sums up his belief in today's cryo for prostate cancer. Compared to surgery and radiation, cryotherapy is cheap, there is virtually no risk in administering it, and because it is precise, predictable and effective its potential benefits include high success, low side effects and a comparatively fast, easy recovery time.

So far you have seen that:
- New percutaneous techniques are **replacing major surgery**.
- **HIFU** (high intensity focused ultrasound) is not yet approved for prostate tumors.
- **RF** (radio frequency) destroys collagen, lacks visual precision, and is size-limited.
- **Cryo** poses less risk to surrounding tissue than RF and has a less painful recovery.

How successful is today's prostate cryoablation?

Many urologists, radiologists and oncologists mistakenly believe that prostate cryo is experimental. Wrong! While we treated the first patient in 1990, in 1999 it was approved for Medicare reimbursement in response to data showing that it is effective and safe. Long-term studies continue to confirm this. For example, in August 2002 **Dr. Duke Bahn** and colleagues published a 7-year study of 590 consecutive patients who underwent cryo as *primary therapy* (no prior gland treatment such as radiation or cryosurgery). They used various success measures:

There is no established uncontroversial definition of successful outcome for cryoablation or for any other treatment of localized prostate cancer. PSA level cutoffs of 0.2 to 0.4 ng/mL are often used in studies of radical surgery, whereas a PSA cutoff of 1.0 ng/mL is more often used in radiotherapy trials. The ASTRO definition of biochemical relapse, 3 consecutive elevations in PSA, is another widely used measure of radiotherapy treatment outcome. The ASTRO [American Society for Therapeutic Radiation and Oncology] definition or a PSA cutoff of 1.0 ng/mL definition of biochemical relapse is, perhaps, the most reasonable measure in trials involving cryoablation therapy. Similar to radiotherapy, prostatic tissue that may be PSA-producing is left intact. This is in contrast to radical surgery, where the entire gland is removed, and with it, all PSA-producing tissue. In cryoablation, there is some preservation of tissue surrounding the urethra…Thus, a PSA level of 0.5 ng/mL…is reasonable for cryoablation.

The use of multiple biopsy results after cryoablation provides an accurate appraisal of local control of cancer, which is the goal of cryoablation therapy. There are several reasons why PSA readings may be elevated despite multiple negative biopsy results, including (1) preservation of residual normal glandular tissue, (2) incomplete ablation of cancer, (3) distant metastasis, or (4) a combination of these.[5]

---

[5] Bahn et al, Urology 60 *Supplement 2A*, Aug. 2002, 7-8

The authors don't compare RP and cryo results because a detectable level of PSA may always be present after cryo but should *never* be after RP. One way they reported their results uses the biochemical or PSA standard of the American Society for Therapeutic Radiation and Oncology (ASTRO). They chose this standard because both radiation and cryo can leave normal prostate tissue in the body that can generate measurable PSA even though the cancer is dead. When PSA stays very low and stable after treatment, the patient is said to be *biochemically disease-free.* Therefore the authors compared their cryo biochemical disease-free results with those of radiation. They also reported follow-up biopsy data.

Prostate cryoablation disease-free success (Bahn et al)

Success rate	Low-risk[1]	Moderate-risk[2]	High-risk[3]
ASTRO	92%	89%	89%
Biopsy-proven	91.7%	84.5%	80.0%

[1]Low risk: Gleason 3-6, PSA < 10 and Stage T2a or less
[2]Moderate risk: Only one of the following: Gleason >6, PSA > 10, Stage T2B or higher
[3]High risk: Two or more of the following: Gleason >6, PSA > 10, Stage T2B or higher

This table shows that for early stage, low-risk patients cryo is a treatment choice with results in the expected success range of RP and radiation therapies. Put another way, the earlier prostate cancer is detected, the more treatment choices a patient has. However, cryo has relatively higher success rates than RP and radiation for men with more advanced localized prostate cancer. The greater the chance of extracapsular escape, the more important it is to get to a cryosurgeon. (To find the odds of escape go to www.prostatecalculator.org and enter age, PSA, Gleason grade and stage.) Remember: ice does not care about aggression. It kills all cell lines, regardless of risk level. The following graph shows a range of published success rates of five treatments in high-risk cases:

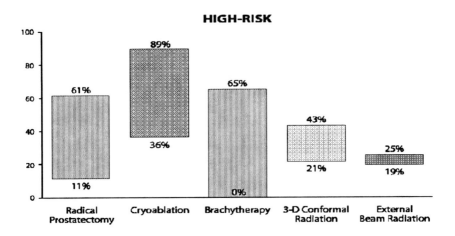

For low risk disease, all treatments are about equal in success, which is why we say that early detection offers the most options, allowing for many considerations like patient preference, other medical conditions, etc. For moderate risk disease, cryo begins to demonstrate statistical advantages as far as lower rates of recurrence. For high-risk disease, however, cryo stands out head-and-shoulders because of its proven destructive power regardless of cell aggression, and because of the size and shape of the treatment area that is possible with ice.

Finally, I often point out that Bahn's work illustrates an important aspect of freezing. Like all prostate cancer therapies there is no guarantee of cure. However, if the cryo missed any cancer it can be detected early. The first PSA taken after the procedure will indicate whether or not the treatment was a complete success. A successful cryo stays successful, unlike other procedures with increasing failure rates as time goes by. Why are success rates stable over time? I have three hypotheses: a) cryo offers immediate **better local control**, by encompassing cancer that has locally penetrated the capsule of the gland; b) early failure can be **readily corrected** by a second treatment; or c) the immune system has been mobilized to make an end of cancer cells that already existed beyond the frozen area (**cryo-immune response**). I have a good feeling that the 7-year data won't be different from the 10-year data or even longer. For me, the revelation that the iceball's power lasts over the long haul is compelling evidence that if a unifocal tumor is encompassed in lethal ice it has the greatest likelihood of a permanent solution.

A herald of new hope

One of the world's largest annual medical conventions meets in Chicago in late autumn. It is the Radiology Society of North American (RSNA), attended by 60,000 radiology professionals from all over the world! Besides the exhibits of dazzling imaging and treatment wonders, some of the world's most brilliant physicians present clinical research papers. Courses are offered for physicians' continuing education. A press corps has special conference access; its members are eager for a scoop.

At the November 2001 RSNA, my co-authors and I announced our preliminary study of nine patients whose prostate cancer had been treated by a focal freeze. The story received so much media attention that Jay Leno joked about it on national TV. In 150 years, ice had gone from a glass laboratory jar, through operating rooms, and onto national airwaves! In March 2004 we presented an update of our research at the Society of Uroradiology meeting in Tucson AZ and to our surprise and delight, our study was judged the best of the conference.

Since that honor, the "Male Lumpectomy" has made major progress toward entering the collective consciousness of the academic leaders who set the agenda for prostate cancer research and treatment. On February 24, 2006 a conference on focal therapy for prostate cancer was held at Celebration Health/Florida Hospital. Represented were many of the major prostate cancer programs in the country including MD Anderson Cancer Center, Washington University, Columbia University and the University of Virginia to name a few. In the morning, research pertaining to focal therapy was presented and then in the afternoon the participants separated into groups to create guidelines to be suggested and adopted by the entire meeting pertaining to the various issues related to focal therapy. By the end of the day the participants had agreed that focal therapy was a reasonable and necessary approach to treating prostate cancer, and that some form of mapping biopsy was essential in the proper staging of patients for a focal therapy approach. The research presented and the consensus statements agreed upon during that conference are soon to be published in a special supplement of the major medical journal _Urology_.

Perhaps the time has come for Dr. Goodbuddy to revisit cryo. "Roger, about that small hard spot on your prostate— your biopsy results came back positive. You have prostate cancer. But don't worry. Your slides show that it's a very small tumor, with no evidence of cancer anywhere else in the gland. Let's talk about all your treatment options, including one I think you're especially going to want to consider." For the first time in the history of prostate cancer treatment, Roger may have the option of killing the whole cancer without killing the whole gland, and thus preserve his fulfilling lifestyle with Rosie.

So far you have seen that:

- Medicare approval means that cryoablation is **no longer "experimental."**
- Long-term data demonstrates **compelling success and safety** with ice.
- ASTRO criteria reveals cryoablation is **at least as effective** as all radiotherapies.
- Biopsy criteria reveals cryo's **distinct advantages** for moderate-to-high risk tumors.
- A successful treatment by cryoablation **stays successful** over time.
- In November 2001 we announced focal freeze results, and opened **new hope.**
- The concept of a "Male Lumpectomy" is **entering the mainstream** of prostate cancer treatment concepts.

Men's Voices

I was familiar with cryogenics from my 28 years as an engineer. The technical aspects as applied to cryosurgery were very convincing to me:

-Cryogenic temperatures kill, on contact, the malignant (and other) cells.

-Cryogenic argon, the medium now used, is easy to control and monitor.

-Ultrasound imaging monitors the formation of the lethal "Ice Ball" to insure that tissues adjacent to the prostate are not frozen.
-Ultrasound imaging provides visibility for accurately placing probes.
-Use of Cryogenics is not new in the medical world, including veterinary medicine.
-Treating growths on or just under the surface of the skin cryogenically is a pretty low-tech "spray can" application.
-More recently cryogenics have also been used to treat internal cancer problems – liver, lung, brain, prostate. Breast cancer treatment is still in the developmental/approval stage. Cryosurgery is a very high tech application.

I cancelled my appointment with the oncologist... Now, five years later, having experienced the operation and the recuperation, I would make the same choice over again. (LJ, Florida)

I didn't want radiation because there's recurrences and complications, which I'd seen in my own practice (for example, rectal and bladder complications). And I didn't like the fact that the radiation treatment would take a long time [to complete]. As a surgeon, of course, I wasn't anti-surgery, but cryosurgery is much easier on the body than either surgery or radiation. (RV, Retired surgeon, California)

Summing It Up

There is a fable by Aesop about a thirsty crow whose beak is not long enough to reach a small amount of water in the bottom of a jug. The crafty bird drops pebbles one by one into the jug, eventually raising the water level high enough to satisfy its thirst. The famous moral of the story is, "Necessity is the mother of invention."

Aesop might well have been describing the history of cryoablation. Perhaps you never heard of Faraday, Arnott, White, Pusey, Cooper, Barken, or Rubinsky. Their individual innovations are like pebbles that raised the level of the resource pool so we could dip into it to satisfy the thirst for a middle ground in prostate cancer therapy.

I am not alone in exploring new applications and ways to do cryo—I stand on the shoulders of giants—and I am most certainly not alone in tailoring a less aggressive freeze to meet the needs of qualified patients. A recent private survey showed that over 60% of prostate cryosurgeons offer safe ways to reduce the extent of the freeze in order to preserve potency for qualified patients. I am, however, the first to follow patients over the long term and publish my data. I have ventured into this new arena—treating unifocal prostate cancer—despite established assumptions and beliefs. The results are compelling, the research is award-winning, and our patients are happy. As with breast lumpectomy and all the patients and practitioners who partnered in the quest, this is how medical history is made.

Remember the clever fiddler at the end of Chapter Two? We have found a way around the devil's bargain of prostate cancer therapy. It stands to reason that if no treatment at all can be advocated for some prostate cancer patients, then attempting to destroy just the focus of cancer in the gland could be a viable option as well, constituting an acceptable and logical middle ground. Next you will see that Male Lumpectomy meets my conditions for an ethical, safe and effective way to resolve the dilemma of clobbering prostate cancer without wrecking a man's life. The careful calculations of how much gland to treat, and the ability to do so with ice, is a coming revolution in men's health the same way surgical breast lumpectomy altered medicine for women.

Chapter 4
The Science Behind Male Lumpectomy

Concern for man himself and his fate must always form the chief interest of all technical endeavors...in order that the creations of our mind shall be a blessing and not a curse... Einstein

History established that ice has the immediate power and adaptability to defeat many kinds of cancer. Cryotherapy had to earn its stripes as all new clinical approaches do. They must be based on sound principles and tested over time. Medical benefits to humankind originate with people who apply mind and heart to serve the highest good.

Before delving into how ice can be legitimately and safely adapted to partial gland treatment, let me tell you some of my personal story that will help in your understanding of how the concept of a Male Lumpectomy came about.

From liver cancer...

As I placed my hand through the incision into Marsha's abdomen to examine her liver using an ultrasound transducer the irony of the situation struck me. The last thing I wanted to be when I was in medical school was a surgeon and here I was, a radiologist, about to attempt an operation that was not only *extremely* dangerous but had never before been performed. It was Marsha's second operation in a little more than a month. In the first, a well known liver surgeon at one of the world's most famous cancer hospitals in New York City had "opened and closed" her after determining that the seven tumors in her liver could not be surgically removed. That had literally been a death sentence for the young mother who had a husband, infant and toddler, waiting to find out the results of what would surely be the last attempt to save her life.

Marsha and her young family had found me through a radiologist in her upstate New York hometown. He had seen me give a presentation on research we had carried out on a new approach to treating liver tumors that were unresectable (could not be cut out). The idea was to use ultrasound to visualize the tumors in her liver, guide freezing probes into the tumors--while avoiding all the major blood vessels--and then destroy the malignancies.

Since it had been established that removing Marsha's tumors was not an option, we were now proposing to kill the tumors by freezing them and leaving the dead tissue in place to be slowly absorbed by her body's defense mechanisms. In the last chapter I mentioned engineer Boris Rubinsky who worked with me on the technical and scientific aspects of this new cancer treatment for over five years. Marsha was literally the first patient with this number, size and kind of liver tumor to be considered for this treatment. One crucial question was truly frightening. After destroying all the tumors, each with a margin of surrounding normal liver, would we leave Marsha with enough functional liver to sustain her life while her liver regenerated after the freeze?

I've been told that the empathy I naturally feel for patients, the ability to "walk a mile in their shoes," is one of my best qualities as a physician. On the other hand, every surgeon must learn to develop a certain amount of detachment in order to assess the safety and appropriateness of each operation. An old surgical adage goes, "Good judgment comes from experience and experience comes from bad judgment." Trained as a radiologist and not having gone through the usual 5-7 years of surgical residency that forges such good clinical judgment, I had to wonder as I started Marsha's procedure whether that same empathy that my patients could sense, was in this case clouding my discernment, making me attempt an operation that was not possible and would end in disaster.

As a child I was not particularly interested in building things. My Erector Set sat unused in a corner of my room. I didn't come from a family where the dad was always fixing household items, nor did I have a big brother who would tinker with his car in the driveway. I grew up in Brooklyn NY and spent my formative years in an apartment where we called the "super" when breakdowns occurred. Mechanical things therefore had a mysterious quality for me. It was not until I went off to college that tools and I made our first acquaintance. I joined the Harvard whitewater club and built my own kayak, my first attempt at anything that was remotely technical or creative.

Later, when I entered my radiology residency, I recognized that I had the ability to view two-dimensional pictures and create three dimensions in my mind. During my second year in residency, I had a revelation that I could use this radiological talent to guide a new kind of "surgery." I was doing a liver biopsy guided by computerized tomography (CT scan) when a serendipitous leap of imagination occurred. I realized that if I could use CT to place a biopsy needle into a tumor, perhaps I could use imaging to deliver something lethal into it, killing it without having to remove it! The advantages were almost immediately apparent. It would a) save normal tissue that usually has to be removed with a tumor, b) create the possibility of treating multiple tumors at the same time, and c) allow destruction of cancer that was sitting on or too close to major blood vessels and could not be surgically removed.

My professors encouraged me to pursue the concept, which led to a National Cancer Institute fellowship and many orders of beef liver from a puzzled local butcher ("Aren't you sick of that stuff?"). We experimented with numerous destructive modalities until we found we could use ultrasound to actually "see" the freezing effect of liquid nitrogen on liver tissue. This visualization was the critical ingredient we had been looking for. We now had a procedure that would allow us to visualize the tumors and surrounding vascular structures, guide freezing probes into them, and then monitor the iceball to confirm that the tumors had been adequately treated. What followed was a long collaboration with Boris to develop equipment and techniques that could be used in patients.

It was this background that brought me to the day and treatment of a person who had been instructed to get her affairs in order. The best medical experts in the world said she had no hope of living. Not without some sweat and trepidation we successfully completed the procedure on Marsha, which she survived without going into liver failure. It appeared my late-blooming surgical judgment was correct! One of the highlights of my career was attending her five year survival party where we celebrated with her family the remarkable circumstances that had brought us together. Today Marsha remains cancer free 15 years after her original liver cryosurgery.

It was clear from this experience that an image-guided approach to treating cancer was going to make a major impact in oncology (the branch of medicine dealing with malignancies). Our newborn contribution was later acknowledged in a letter written by Dr. Susan Wall, then President of the Society of Gastrointestinal Radiologists, nominating Boris and me for the Russ Prize, the highest honor of the National Academy of Engineering:

> As President of the Society of Gastrointestinal Radiology it is particularly appropriate that I write this letter since our society was one of the first to recognize the enormous potential of their work when we awarded them the first ever Roscoe Miller Award in 1986. I am proud that our society had the judgment and foresight to bestow our award on research that would later prove to have such a major impact on the direction of oncology care and research. Imaging guided cryosurgery and the other imaging guided destructive modalities such as RF, laser, microwave, that grew from the success of their work clearly represent even today the only other potentially curative treatment other than resection for liver cancer. While initially started in the liver their concepts are now being successfully applied in other tumor systems notably the prostate, breast and kidney.

...To prostate cancer

We knew we were onto something important. Once I saw the liver cancer results I wondered, what's the next good target? Compared to the liver, I felt that treating the prostate would be technically even easier. I did not anticipate the challenges I would encounter in this new endeavor.

I reviewed the literature on prostate cancer therapies, especially RP, and found at that time a 30-40% reported rate of positive margins (leaving some tumor behind in the prostate bed). Perhaps because I had not come up through urology training, I was more open about other modalities besides surgery. I wondered if I could improve on the statistics by treating prostate cancer with cryo. I was an Associate Professor of Radiology and Neurosurgery at the University of Pittsburgh at the time and when I approached the Chairman of the Department of Urology with the idea of adapting my liver cancer work to the prostate, I met stiff resistance. It became clear that if any progress was to be made in this arena, I was going to have to leave my job at the university and find a urologist to work with me. I found a willing partner across town in one of the young urologists at Allegheny General Hospital, the institution where I did much of my early liver work. When a human experimentation committee approval was obtained for our initial protocol I made the move and left my job at the university.

Our goal at the time was to freeze and destroy the whole prostate, essentially getting the same results as a radical prostatectomy but without surgical removal. We were driven to do this by two factors: first, the patients that needed the most help were already poor candidates for surgery due to their extensive and aggressive cancers (as you saw from Bahn's data, high risk patients are ideal candidates for cryo); and second, we were still influenced by the erroneous assumption that prostate cancer is always multifocal, requiring a radical whole gland surgical removal.

As I gained experience with the full spectrum of prostate cancer patients, however, I became curious and was prompted to look into the pathology literature to reexamine the issue of the multifocality of prostate cancer. To my surprise I found data that many prostate cancers were not multifocal, suggesting they might not need an aggressive whole gland treatment. At first I was surprised that this had been overlooked or ignored, but then it seemed to make sense. If none of the available treatment approaches could effectively treat less than 100% of the gland, it would make no difference if a patient's disease was multifocal or unifocal, right? I had to think this through to my own satisfaction.

The following illustration is an artist's conception of how a tumor might grow in only one location. In my mind's eye I could imagine how I could encompass such a tumor on one side of the gland in a ball of ice yet spare the other side.

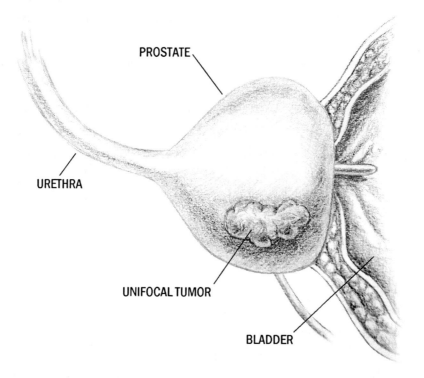

PROSTATE

URETHRA

UNIFOCAL TUMOR

BLADDER

A tumor in only one location (unifocal disease).

Reexamining RP I asked, can we **surgically remove** only part of the prostate? Theoretically yes, but logically who would go through major surgery yet leave part of the gland behind? It's hard on the body and recovery can be long. Once a doctor has gone that far—in effect, a major pelvic invasion—why leave anything behind? I also didn't think that removing part of the gland would improve the positive margin rate either, but would probably make it worse; best to remove the entire gland and be done with it.

I then pondered whether any form of **radiation** could deliver a truly focal cancer kill. Radiation scatters in all directions. Even a partial seed implant would radiate the other side of the gland, the bladder, rectum and nerve bundles. There is no way to shield areas where radiation is not needed. A second risk of radiation exposure is that it may cause insignificant amounts of cancer elsewhere that are not killed to grow and mutate. A treatment in one location might thus cause worse disease in another. Finally, once radiation damages tissue around it, salvage options become fewer and more difficult, with higher rates of undesirable side effects.

Both surgery and radiation were impractical and illogical for partial gland treatment. So that left cryo. I believe that we applied good scientific principles and our own innovations to the logical standards we established for a partial freeze. Ethically, how could I suggest an untested treatment to a patient? Here is where serendipity came in. The patients selected themselves, as the following story illustrates.

One day a patient came to see me. He was young, and had a well-defined tumor. He had been on watchful waiting for about a year, but was not comfortable with waiting any more. Like so many women with breast cancer, he asked, "Can't you deal with just the tumor?" We had this untested idea—focal freeze of the prostate—and the patient was willing to take the chance once he learned he apparently had nothing to lose because so many options would remain open to him if it didn't work. But work it did! It is now over twelve years since his treatment; he was continent and potent immediately after the procedure and remains so. Best of all, today he is cancer-free!

At first the news traveled at a crawl. We didn't publish or advertise the results because our patients were too few and our follow-up too short. I didn't want to get patients enthused about a new controversial option for prostate cancer and possibly treat large numbers of them before we had a good sense of whether we were correct in our thinking. Yet many early patients still sought me out. Those who located me were very determined men. They had to look hard, partly because I'm not a urologist. They were a unique class: physicians, engineers, executives. Some had research backgrounds, others were used to getting their way and not deterred by obstacles. They were pioneers with us in staking out new medical territory, like the women with breast cancer who sought out breast lumpectomy against all medical advice. Because of their tenacity and commitment, and our mutual trust, I was determined to establish the credibility of Male Lumpectomy. I needed a yardstick by which to measure it. I developed ten standards I felt it must meet.

Let's examine the feasibility of the Male Lumpectomy using these ten criteria.

#1. Is Cryoablation A Feasible Way To Kill Prostate Cancer?

As you read in Chapter 3, ice has a history as an accomplished cancer assassin. I knew from history that ice is effectively lethal, with proven success in treating men with prostate cancer. The type and aggressiveness of the cell line does not matter.

#2. Is The Technology Time-Tested And Available?

The newest generations of cryotherapy technology for prostate cancer are remarkable systems that literally integrate the genius of doctors like Israel Barken, Fred Lee, Douglass Chinn, Boris Rubinsky and others. Their long experience and careful research led to key improvements. Their collective contribution is exemplified in today's advanced cryosurgical systems.

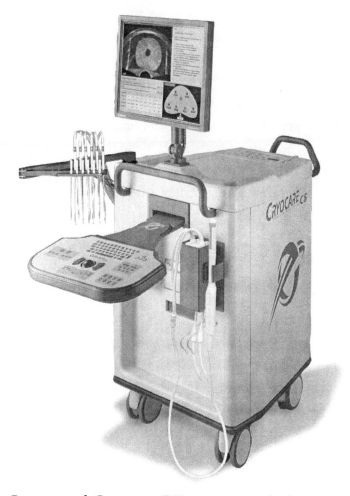

Integrated CryocareCS® system includes computer mapping of gland and probe placement, urethral warmer, cryo and temperature probes and real-time ultrasound guidance.

Refined ultrasound guidance, computerized algorithms for properly sized, shaped and cooled iceballs, thermal feedback, warming catheter, user-friendly monitor screens, an optional grid to guide probe placement, and adaptability for each cryosurgeon's professional judgments assure accuracy and safety for each patient. The system we use generates the same length iceball used to gather the long-term data. It applies **six to eight probes for a total gland freeze**, depending on the size of the gland and the extent of the area to be frozen. With fewer probes this same equipment can **pinpoint a smaller freeze area**, up to almost the entire gland, creating the equivalent of a lumpectomy. It destroys a focused area instead of the whole gland. The technology is available, and has demonstrated consistently high success in treating prostate cancer.

#3. Can Focal Freeze Offer A Treatment Safety Margin?

No one wants to go through prostate cancer treatment if he thinks living cancer is still left in his body. An important issue regarding focal freeze is whether it can deliver a **safety net** against possible spread. Look at the illustration.

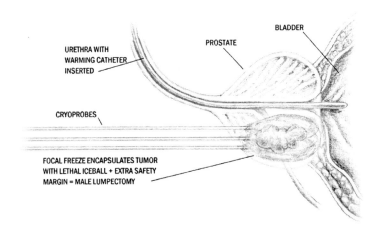

The finished iceball extends beyond the tumor on all sides.

The technology makes it possible to customize the size and shape of the iceball for treatment beyond the known tumor location. The Male Lumpectomy extends ice into the areas most vulnerable to cancer invasion for each patient. I treat every patient **as if his local cancer has begun to spread.** For example, if a moderate risk tumor is toward the gland posterior (rear) on the left, I will shape the ice into the prostate bed and seminal vesicles on that side by placing cryoprobes there. This short-circuits possible local extension of cancer outside the gland (spread). I routinely freeze the neurovascular bundle on the side of the malignancy to prevent any possible cancer spread along the neural "superhighway" if it penetrated the nerve sheath (again, pushed by patients who wanted both neurovascular bundles spared, we are carefully investigating this possibility—so far, so good!). However, the non-frozen nerves on the other side can still function to allow erection. In most patients I also treat the area where the seminal vesicles adjoin the gland. This is significant, because in our published studies **45% of patients had moderate to high risk** unifocal disease! It demonstrates that a partial freeze can still be a treatment option for men with a Gleason grade of 7 or 8, provided their disease is unifocal. For some patients, the extra-thorough safety margin may temporarily diminish erectile function because the pelvic bed has dense blood vessel structures (vascularity) that could be affected by an aggressive partial freeze, so some of my patients can experience a delay in the return of sexual function. However, their faith that they are cancer-free more than compensates for a few months of inventive sex while waiting for potency to return.

Is it certain that the cancer is destroyed? Remember that no cancer treatment comes with a guarantee. Our results, though, are as promising as early breast lumpectomy was for women because we take no chances. This is a **minimal yet radical** treatment. The treatment area is reduced but includes a precautionary extra margin of kill. Male Lumpectomy is a combination of focal aggressive ice with a safety margin.

#4. Is There A Way To Identify Appropriate Candidates?

The main objection to a Male Lumpectomy is based on the hypothesis that all prostate cancer is multi-focal, in which case no one would be appropriate for less than a total-gland freeze. As in breast cancer, prostate cancer is a spectrum of different cell lines. Some may be treatable by lumpectomy and some may not. Since the PSA blood test, prostate cancer can be found very early when the greatest number of options exists. If the tumor is "clinically significant" based on its volume, aggressiveness, etc., the doctor will probably encourage treating it. In all likelihood, his treatment recommendation will be a whole-gland approach based on the assumption that where there's one tumor, there must be others, no matter how small and undetectable.

Multiple pathology studies in the literature looking at prostate glands removed surgically then examined, clearly show that about **one third of clinically significant prostate tumors appear to be unifocal.** Over 220,000 men are diagnosed annually with prostate cancer. This means that at least **70,000 men might qualify for Male Lumpectomy.** Based on tumor clinical significance, the actual number of potential candidates may be even higher. Up to 80% of multi-focal tumors appear clinically insignificant, meaning they probably do not need immediate treatment because they are so small. With thorough biopsy analysis, perhaps more than 100,000 men can consider this treatment option. They fall into two categories:

1. Those men for whom watchful waiting is a treatment option, and

2. Those men with apparent unifocal, unilateral (one side of the gland) or clinically insignificant multifocal disease.

For those in the first, I would say, "If you're willing to do nothing, then it seems to me you should have an option in between without the morbidity of whole-gland treatment." For those in the second, our results speak for themselves.

When I meet with a man whose diagnosis places him in one of those two groups, I further qualify him with **an extensive prostate biopsy.** My name for this type of biopsy is perhaps more descriptive: **3D Prostate Mapping Biopsy (3D-PMB).** Remember those red/green cardboard movie glasses when you were a kid? They made you see the film as if it occurred in three dimensions or "3D." A 3D-PMB is extremely precise in three dimensions throughout the length, width and height of the prostate. Unlike the usual trans-rectal biopsy, usually 8-12 needles, this procedure is done through the perineum (skin between the scrotum and the rectum), the same area that is penetrated for brachy or cryo. While the patient is sedated, I use a grid to guide this methodic procedure, taking needle samples five millimeters apart, sometimes 30-50 samples. I formerly concentrated on the non-cancerous side of the gland, searching for clinically significant cancers, but as the safety of the procedure has been confirmed we have now extended the biopsy to the whole gland. My 3D-PMB data shows that 50% of patients will have cancer demonstrated on the side of a previous negative TRUS biopsy. Our thinking is evolving a little on this matter, however. The 3D-PMB gives such a precise idea of the cancer location, if a positive biopsy is obtained but the tumor is not near the neurovascular bundle then that area of the prostate can still be spared and the goals of a lumpectomy still realized. Even patients with two tumors demonstrated can be treated with a focal freeze depending on the size and location of the involved areas!

So far you have seen that:
- **Neither surgery nor radiation** meets our criteria for prostate lumpectomy.
- Cryoablation is a **proven** way to kill cancer.
- Using current technology, a focal tumor can be **killed by focal ice.**
- An extra iceball **safety margin** also destroys microscopic cells at risk of spreading.
- Men on **watchful waiting** should consider the option of Male Lumpectomy
- Men with **unifocal or unilateral disease** should consider this option as well.
- Men with **clinically insignificant multi-focal disease** should also consider it.
- Patients are qualified for this approach by using a **3D Prostate Mapping Biopsy.**

#5. Have Early Male Lumpectomies Been Successful In Cancer Control?

The results in our first published studies are compelling. At the RSNA convention in 2001, we announced a small preliminary study with nine patients who had a focal freeze that spared one nerve bundle. Follow-up ranged from 6 to 72 months (average 36 months). We reported that all patients were biochemically disease-free (stable PSA). Six patients who were routinely biopsied were negative for cancer.

At the 2004 Society of Uroradiology, I updated this earlier work. Of 21 patients who underwent focal cryotherapy for their prostate cancer, 95% showed no biochemical evidence of cancer recurrence (stable PSA). None of the 17 patients who had biopsies after therapy showed any evidence of cancer. What's significant is that the average follow-up in these patients was 4 years and that half of them were of medium and high risk for recurrence and would therefore not have been candidates for seed implantation. Also of great importance is that these results were achieved without additional radiation therapy as is needed in breast cancer lumpectomy.

The one patient (who also was a physician) in this series with an unstable PSA and therefore recurrent cancer, illustrates an interesting point that often isn't considered in prostate cancer treatment. This patient had a Gleason 8 cancer, a PSA of 12 and was a stage T2b. In other words he had every risk factor placing him in the category for high risk of cancer recurrence. Since his cancer was unilateral we opted to treat him with focal freezing. He was continent and potent immediately after the procedure. When his PSA did not stabilize after cryo he had an extensive 3D-PMB with all biopsies in both the frozen and unfrozen areas showing no evidence of cancer. It shortly became clear that he had metastatic prostate cancer. In a paradoxical way this patient also illustrates the success of the Male Lumpectomy. His focal freeze provided excellent local control of a very aggressive tumor yet he was exposed to minimal morbidity in gaining that result. What could be worse than treating a patient with a morbid whole gland treatment, ruining his lifestyle with major complications such as incontinence—and then learning it was all for naught since the cancer had already invaded his bones or lymph nodes?

In December of 2006 I was invited to speak at the annual meeting of the Society of Urologic Oncology held at the National Institutes of Health, to present our latest data on the Male Lumpectomy. There were more patients and longer follow-up but the results had not changed over time. We still had 95% of our patients who had no evidence for persistent cancer based on their PSA results. This was also the first presentation where I proposed a rather revolutionary concept based on our data "that a focal treatment may actually provide **better** cancer control than radical whole gland treatments!"

Let me explain. When we first started treating patients with focal therapy we thought we would be trading off the lower morbidity (complications) associated with such a treatment with an expected higher local cancer recurrence rate. In fact the data suggests otherwise; it now appears that the local control of prostate cancer is actually **better** with a focal treatment than a whole gland treatment!

At that same meeting, to my chagrin I had to present the fact that of the 55 patients we had treated by focal freezing, with an average of 3 years follow-up, we had not a *single* patient with evidence for persistent local prostate cancer. Why to my chagrin, you might say? Well, the desire of any medical researcher is that his results be believable and reproducible. Here I had just presented data that no other prostate cancer treatment had ever come close to achieving in a similar patient population (remember that 50% of the patients were in a medium to high risk category). I had just set the bar so high that reproducibility would be very difficult to achieve. Had my patients been just statistics instead of real people with loving families, I might have hoped for at least a few "failures" right about then!! Luckily, similar results are already being reported in the literature by at least one other institution, in turn giving our results greater credibility (Bahn et al, *Journal of Endourology*, Sep. 2006[6]).

The reason that a focal therapy should provide better cancer control than whole gland treatments is very understandable. I now believe that whole gland treatments compromise killing or removing the cancer by striving to treat the whole gland unnecessarily and by disregarding the location and the extent of the cancer when planning the treatment. For instance, rather than concentrating higher doses of radiation therapy in the area of cancer to insure its destruction, radiation is now spread over the whole gland. The dosage of radiation has to be limited to prevent damage to surrounding normal structures such as the rectum and the bladder. In focal therapy, on the other hand, the location of the cancer is known and a destructive freeze can be delivered precisely to the tumor and its success can be confirmed using ultrasound and temperature monitoring at the time of the treatment.

---

[6] The study consisted of 31 men with a mean age of 63 who underwent the focal cryoablation procedure and whose disease was believed, through targeted and systematic biopsies, to be unilateral or confined to one sector of the prostate gland. At a mean follow-up of 70 months, 92.8 percent (26 of 28) remained biochemically disease-free and 96 percent (24 of 25) had no evidence of cancer on post-treatment biopsy. The one biopsy-positive patient had his prostate subsequently treated with full gland cryoablation and currently remains biochemically disease-free. Follow-up for the study consisted of PSA measurement every three months for one year and every six months thereafter, with biopsies at six months, one year, two years, five years and following any three consecutive PSA rises.

Our statistics reflect the disease-free success of Male Lumpectomy. I expect them to hold over time just as the results for whole gland cryotherapy have. For now I am completely confident in the ability of a focal freeze to destroy a unifocal tumor with at least the same success as RP or radiation.

#6. Can Focal Freeze Truly Spare Sexual And Bladder Function?

Using the same studies, let's look at our results regarding post-treatment sexual and bladder function. In the 2001 study, 78% of men maintained potency to their satisfaction; by 2004 the number had risen to 85%, a remarkably consistent record for local gland treatment. Unlike RP or total cryo, post-treatment orgasms are not necessarily "dry." In many patients the spared portion of the gland will still produce prostatic fluid. As for urinary control, all patients without a history of previous prostate surgery have been 100% continent immediately following treatment, defined as *never wearing any pads*—the strictest standard for measuring continence.

The challenge of localized treatment is to be like a "smart bomb" that kills the enemy without harming the lives of civilians. Male Lumpectomy is "Smart Cryo." No other available treatment is so minimal yet radical.

#7. Will Recovery Progress Rapidly And Without Need For Hormones?

Cryotherapy, even an aggressive whole-gland freeze, usually has a recovery time in the range of a few days to a week before return to normal activity. A total gland freeze also requires the temporary use (1-2 weeks) of a catheter to aid in emptying the bladder because the urethra may slough dead lining cells during the recovery period.

With a partial freeze, **recovery is even quicker.** It's like "Cryo Lite." There is less trauma to the urethra, so catheter use is very brief if at all. Swelling and bruising may still occur, but is quite manageable. Most men report a little discomfort that responds to oral medication if needed. This liberation from a long RP recovery, or weeks of external beam radiation, is a blessing for men who want to return to work, moderate exercise, and social life quickly.

Also, **no hormones** are necessary after treatment. For those patients who were temporarily on pre-treatment hormones to reduce gland volume, once cryo is done hormones are discontinued. More exciting is the implication for thousands of men who are currently on androgen blockade in hopes of deferring or avoiding the morbidities of local treatment. With our approach they could be cancer-free without impairing lifestyle, by eliminating those hormonal side effects.

#8. Does Male Lumpectomy Integrate With Other Therapeutic Approaches?

Cryotherapy integrates very well with complementary medicine (also called alternative or natural medicine). Many cryosurgeons who offer a freeze tailored to less than 100%

of the entire gland encourage such patients to reinforce prostate health the way those on Conservative Management do. Regular exercise, reduced dietary fat, vitamins and other nutritional supplements boost the immune system. Common techniques of stress reduction, including meditation, also enhance well-being.

The literature notes that a *cryoimmune response* has been observed in humans as well as in lab animals. There are varied theories about the mechanisms involved, and research continues. After cryo the immune system may do a "seek and destroy" mission, attacking any of the same cancer cells that might have begun to spread elsewhere. This appears to weaken and possibly destroy them. Where radiation may cause surviving cells to become more aggressive, cryo may make them less aggressive. In medical terms, they are then *more amenable* to other treatments such as chemotherapy. However, this cryoimmune response is not universal, and there is no way to predict which patients may have it. And, as you will shortly see, focal freeze does not burn any subsequent treatment bridges in the event of recurrence.

#9. Can Possible Recurrence Be Reliably Monitored?

The same tests used to keep an eye on all post-treatment patients (PSA, free PSA, ultrasound, color Doppler ultrasound, MRI spectroscopy) are used for reliable supervision following a tailored freeze. Unlike total gland treatments, Male Lumpectomy intentionally spares living prostatic tissue. Hence, PSA level is expected to be detectable but low and stable. Following treatment, I monitor a patient's PSA closely for two years (every 3 months), since treatment failure after cryo usually shows up early. We used to routinely biopsy all patients at one year following the procedure, but when we found that all patients with stable PSAs had negative biopsies, we reserved re-biopsy for special cases. After two years, future monitoring may be less intensive (every 6-12 months). At any sign of rising PSA or suspicious biopsy, further tests determine if localized cancer has recurred. This leads to my final criterion.

#10. If Cancer Recurs, Do Patients Have Treatment Options?

Male Lumpectomy patients have the best of all worlds. If it is successful, their cancer is gone. However, if routine follow-up detects a rising PSA or suspicious biopsy, I can conduct other thorough tests to determine if localized cancer has recurred. A significant advantage of Male Lumpectomy is that the full range of salvage treatment options still remains if cancer recurs: RP, beam radiation, brachytherapy, repeat cryo, even watchful waiting and conservative management can all still be carried out with no added complications. This is why I can confidently state that Male Lumpectomy "burns no bridges."

So far you have seen that:
- **Award-winning research** on Male Lumpectomy indicates excellent cancer control.
- Male Lumpectomy shows **80%-85% success in preserving potency**.
- Male Lumpectomy has **100% success in preserving continence defined as zero pads**.
- **Recovery is rapid**, often requires no catheter, and eliminates the need for hormones.
- Male Lumpectomy is **harmonious** with other treatment approaches.
- **Monitoring for recurrence** is easy, safe and reliable.
- If cancer recurs locally, all treatment options are **still available**.
- Male Lumpectomy **meets all criteria** for a happy medium in treating prostate cancer.

Men's Voices

My wife and a close friend who helped me review the information concluded that cryo made a lot of sense for several reasons:

-The procedure is curative in many cases if cancer has not spread outside of the prostate.

-Swift death to the malignancy. Freeze, thaw, freeze, thaw. It's over in less than 2 hours.

-Minimally invasive (via the perineum), no cutting, minimal blood.

-Only two hours of anesthesia, local or general. Outpatient or overnight hospital stay.

-Partial nerve save (one side) possible in some cases.

-No radiation hazard.

- -Improved procedures, high tech monitoring and high tech equipment have all but eliminated unintentional freezing concern.
- -The procedure, unlike the other options, can be repeated if the cancer reoccurs. (LJ, Florida)

When you hear 'gold standard' your mind makes the connection, it kills the cancer. But it's not a foregone conclusion. With all my reading, it was clear that surgery and radiation were not as sure as I had been led to believe, and then you had to put up with the side effects. It wasn't the way I wanted to do things. (LC, New Jersey)

Summing It Up

Over eleven years ago we started to investigate if a focal approach to treating prostate cancer was feasible. Patients were extensively re-biopsied on the side opposite their demonstrated cancer; if no cancer was thus detected, only the side containing the tumor was frozen. The procedure was a uniquely minimal yet aggressive. The neurovascular bundle on the side of the lesion was always treated and in most patients freezing included the confluence of the seminal vesicle (bilateral nerve-sparing is now being carried out in appropriate cases). An additional factor that makes this approach attractive is the fact that the literature shows that cryo can be repeated if needed. Consequently, no bridges are burned in taking this targeted approach.

A lumpectomy approach to treating prostate cancer using cryoablation appears to hold great promise. The middle ground of treating just the cancerous area of the prostate could have the same salutary effects for men that lumpectomy has had for females with breast cancer. The ultimate success of such an approach will hinge on accurately determining the extent of cancer in each particular patient since there is already ample evidence that cryo is an efficacious means of treating prostate cancer. Much needs to be done to fully document the success of this approach, with large patient series followed for long periods, as was accomplished for breast cancer lumpectomy. Based on what we know and the criteria I have outlined, for a patient to enter into a clinical study on Male Lumpectomy would be a reasonable—even rewarding—risk.

Let's see what other professionals say about "Smart Cryo."

Chapter 5
What Do Other Experts Say?

... for destruction ice
Is also great
And would suffice.
Robert Frost

If you've seen a good magic act, you may have experienced the childlike wonder of wanting to believe your eyes, though rationally you know it's a trick. Making something disappear before our eyes appeals to the part of us that wishes miracles were true. Magicians don't repeat tricks—they leave us wishing that magic happened in life.

Science is sometimes serendipity, but it's never magic. It demands repetition. Experiments are carefully controlled and recorded so they can be replicated. Results must be demonstrated over and over. Only when outcomes are consistent under the same conditions are they considered meaningful. Likewise, doctors want to repeat their successes, but it doesn't always work out that way. They want to preserve life and its goodness, but every patient is unique. One man who had decided on cryo joked with his cryosurgeon, "I'm glad I'm not your first cryo patient because I wouldn't want you experimenting on me." His doctor smiled and replied with humble honesty, "All patients are experiments." If this seems blunt, here's an inspiring true story of a prostate cancer patient who saw himself as a living laboratory. A lifelong cancer researcher learned about his prostate cancer too late for primary treatment—it had already metastasized. He went on hormones. Then he also went on a retreat. There he realized that his cancer was an opportunity for inner spiritual vitality, and also a chance to generate data that could serve others. He found a teamwork-oriented oncologist. Together they developed his unique treatment, recording the protocol so others could repeat it if successful. Still alive and energetic, he still uses his own life to contribute research in the service of science.

When a doctor gets the same results with many patients, it raises his confidence. Validation from others' work also matters. If I am the only person in the world who is doing partial freeze, no matter how many of my patients remain cancer-free it is of less value to the rest of medicine if other physicians cannot replicate this approach with similar results. This chapter explores the knowledge and contributions of other experts on focal freeze. After all, it's not magic.

John Rewcastle, Ph.D.

John Rewcastle has been called "Dr. Ice." He's not a physician, but it's an apt nickname for a Canadian raised in arctic winters who also understands cancer death by cold. He holds a doctoral degree in medical physics, having delved into the nature of the physical world from Newton to Einstein. He found his academic niche in the relation between heat and mechanical energy, or thermodynamics. Events led to a career studying what happens to tissue when it's frozen. In short, John Rewcastle is an expert *cryobiologist.*

Rewcastle is **certain that ice kills cancer**. He explains, "There is a very large body of scientific and clinical evidence that if you freeze tissue cold enough it will die. It's common sense. You freeze a climber's toe on Mt. Everest and his toe falls off." He is so sure of the lethality of ice that he states there is no need to reestablish this with laboratory studies in the case of a partial freeze of the prostate. "We know that when you freeze tissue sufficiently you will kill it. There is nothing fundamentally different occurring in this case. Focal cryoablation is simply creating a smaller zone of destruction."

It sounds simple but logically you may wonder, "If ice is so effective, why do they say cryo is repeatable? Why would it be necessary?" Rewcastle points out that in an ideal world, a doctor would know with 100% certitude that he's really freezing what he intends to freeze. However, there is a risk of missing cells, just as there is with surgery and radiation. No one can guarantee that the scalpel or the beams got everything. Since the safety of the patient comes first, the attempt to preserve nearby healthy structures means there's always a chance of not getting some tissue cold enough and therefore not killing it.

If John Rewcastle were a cryosurgeon, would he be willing to perform a partial freeze on a patient? Before responding, Rewcastle poses a challenge: "How do I know he has a unifocal tumor? This is the whole crux of the matter. The question is not whether freezing kills tissue, it is whether or not partial treatment of the prostate will sufficiently treat a man's cancer. There is a patient population that will absolutely benefit from a partial freeze! But how do we know who they are?" He identifies two essential determinants:

1. A saturation biopsy (what I call 3D-PMB) taken with very small increments between samples is necessary for a thorough diagnosis. The efficacy of this approach was confirmed by a recent study published in the _British Journal of Urology_ by Dr. Crawford of The University of Colorado. This study simulated the 3D-PMB

approach on prostates removed at RP and autopsy. It demonstrated that the 3D-PMB biopsy found 95% of the known significant tumors (5mm or greater).[7]

2. Predictors of the probability of unifocal disease based on such things as age, stage, PSA, family history, etc. Research is ongoing to define these predictors.

If the patient has a unifocal lesion, "Dr. Ice" is confident that Male Lumpectomy is a good treatment with a full range of back-up choices.

David Bostwick, M.D.

Dr. David Bostwick is the Director of Bostwick Laboratories. He is nationally known and respected as one of the foremost experts in *urologic pathology*, the study of the origin and nature of prostate disease. He describes himself as a student of the prostate for 25 years, and has systematically and scientifically explored many aspects of prostate cancer. He states, "Based on all of my experience evaluating whole gland prostatectomy specimens there is no doubt that a significant number of patients would be appropriate for a focal treatment of their prostate cancer."

Currently, Dr. Bostwick is a principal investigator in research to determine the statistical occurrence of unifocal prostate cancer. The collaborative protocol involves not only his own facility but also Baylor University, and the universities of Florida and Colorado. Approximately 500

---

[7] Crawford, E.D. et al. "Clinical staging of prostate cancer: a computer-simulated study of transperineal biopsy." *British Journal of Urology.* 2005 Nov;96(7):999-1004.

prostate glands were carefully obtained from patients in several locations. Each gland is serially sectioned, like cutting a loaf of bread in very thin slices, and entirely looked at. In order to correlate all variables for statistical analysis, the data base for all patients includes the usual and customary factors, such as age, Gleason grade, gland volume, surgical margin status, tumor focality and site, etc. This database is evaluated for completeness before it goes to the statistician.

Some men have unilateral (one side) or unifocal (one tumor site) disease. Others have bilateral (both sides) or multifocal (several tumor sites) but Bostwick says that according to many published papers the small cancers may not need immediate treatment. He hopes to use data from his present study to develop a model to predict which patients, given all known factors about them and their prostate cancer, might qualify for a *conservative or minimal treatment* that will not affect quality of life yet have a high degree of treatment success. Current predictive tools, particularly the Partin Tables, are an effort at predicting recurrence and survival, but can't always offer the best guide to treatment. As a result, intervention is often overkill—unwarranted total removal or destruction, with needless risk to continence and potency. If the tumor can be accurately identified, then theoretically therapy can be tailored to each patient if enough is known about his disease. Dr. Bostwick is excited about the 3D Prostate Mapping Biopsy we are doing since it helps give specific definition to unilateral or unifocal tumors.

Before today's advanced cryotherapy there wasn't a technology that could truly target only the tumor, or a cancerous portion of the gland close to total but sparing erectile function. Freezing is specific enough to destroy a tumor as small as the head of a pin, though Bostwick states that no one uses it that way. Instead, most doctors who do less than a total gland freeze still destroy a moderately large additional area. He says, "Of all the ablation techniques, cryo is probably the most useful because it's controllable and confirmable at the time of the procedure." The conformal quality of ice means it has an important role in the future of treatment.

Daniel Rukstalis, M.D.

Speaking of conformal, Dr. Dan Rukstalis, (Director, Department of Urology, Geisinger Medical Center, Danville, PA), has been on the cutting edge of treatment since he began seeing prostate cancer patients. Rukstalis observes that there are now so many good therapies that men can and should make individualized choices. He develops teamwork with each patient. He gets to know not only each man's clinical parameters, but also how he sees his life and his future. He embraces a philosophy of choice, committed to patient education and empowerment. He admits, "In a way it's selfish, because it's how I would want to be treated."

Rukstalis was first attracted to the promise of cryo over a decade ago. "I was an early adopter and developed new applications. I was among the first in the country to do renal [kidney] cryo." He co-founded the first kidney cryo program at the Medical College of Pennsylvania. Besides prostate and kidney cryo, he also offers adrenal cryotherapy.

From the beginning it was clear to Rukstalis that many men can live a long time without having their disease treated. He wrestled with the notion that some patients were probably over-treated. He encouraged men with low-risk disease to wait, monitor, and change their lifestyle, but many were uncomfortable and wanted whole-gland treatment. Their fear of eventual cancer spread outweighed the logic that it was probably not life threatening. Was there a balance between providing the reassurance of treatment yet not put continence and potency on the line? Freezing proved to be the answer. "I don't believe we should leave prostate tissue in a man's body untreated for no reason. The only reason for sparing prostate tissue is to reduce incontinence, impotence and urethral sloughing or stricture. With cryo we can kill more than 75% of the prostate and still avoid those side effects." He began offering what he calls *conformal cryoablation*.

Rukstalis does not suggest that cryo is the best or only treatment approach. He has a balanced view of its place among all options. In cases where radical prostatectomy would be an excellent choice, Rukstalis is a skilled surgeon and can perform that operation. On the other hand, he knows that with earlier detection, many prostate malignancies are discovered before reaching a stage where radical or total treatment is indicated. He notes that with the ability to adapt cryo to each patient's circumstances, the earlier you find the cancer the more likelihood that a conformal freeze can be offered.

The principal of conformal cryo is simple: Treat the region(s) at highest risk while sparing areas that would cause the patients side effects if treated. In other words, any fraction of a gland less than 100% can be treated as successfully as the total gland. Rukstalis estimates that 90% of his cryo patients have been treated conformally. All of them are followed and monitored by DRE, serum PSA and repeat biopsy at 6-12 months after treatment; about 15% are found to have a micro-focus. These tiny lesions may be determined to be clinically insignificant. Because cryo integrates well with other treatments, Rukstalis educates his patients on the value of complimentary medicine and lifestyle changes, citing published evidence that healthy nutrition, exercise and stress management may help control prostate cancer.

In the event of recurrence, Rukstalis' conformal patients have all other treatment options available, so again no bridges are burned. They can have RP, brachy or repeat cryoablation as an additional treatment.

Rukstalis is aware that many cryosurgeons are quietly doing less than total gland cryo. They may be comfortable in a private doctor-patient relationship, but it's hard to "go public" in the face of the long-held view of multi-focal disease with each focus being of equal risk. "It's only the last five years that pathology has shown that the tumor volume is going down at time of diagnosis because of earlier detection," says Rukstalis. Ice can be adapted to treat men early in their disease without the current level of morbidities. "This is a safer approach for the patient than total cryo because the decision to modify the treatment in the areas of highest risk reduces the side effects."

So far you have seen that:
- When properly identified, unifocal and unilateral tumors can be killed by ice.
- Earlier diagnosis makes it possible to treat patients in conformity with their needs.
- Conformal therapy is an example of increasing interest in the Male Lumpectomy.

Men's Voices

As a Canadian, trust me, ice kills. (J. Rewcastle, Ph.D.)

What Onik is doing is the future of prostate mapping. (D. Bostwick, M.D.)

The key for me, the wonderful thing about cryoablation, is the ability to tailor it to the individual patient. If the patient and doctor want to treat one spot, they can do that. If they want to treat 100% they can do that. But if they want to treat more than that spot but less than 100% they can do that too. Each individual patient has the ability to be involved in his treatment decision. (D. Rukstalis, M.D.)

Summing It Up

Burger King Corporation began in Miami FL in 1954. Its mission was simple: reasonably priced quality food served quickly in a clean, attractive atmosphere. Burger King now has over 11,400 locations in 57 countries. Mission accomplished!

McDonald's Corporation presents ongoing competition. Despite being the first fast-food chain to offer dining rooms, and the 1975 addition of drive-thru service, Burger King needed a way to distinguish itself from the "Golden Arches." Its first catchy slogan, "The Bigger the Burger the Better the Burger," had a satisfying ring.

There have been many ad campaigns since then, but the one that debuted in 1974 stands out: HAVE IT YOUR WAY®. This registered trademark captures Burger King's vision of customizing meals according to individual taste. Later adaptations—e.g. offering healthier choices for nutrition-conscious customers—are variations on this brilliant idea. Those four words are implicit in subsequent Burger King slogans.

McDonald's is a great fast-food giant, but it's not the only burger in town. The success of Burger King testifies to the appeal of having your order tailored to your own satisfaction. Like a Whopper®, cryo can be customized. Onik's Male Lumpectomy or Rukstalis' patient-specific conformal freeze make it possible for each patient to truly *have it his way!* In this sense, cryo is the "Burger King" of prostate cancer treatment.

Now read about three men who "had it their way."

Chapter 6
What Do Survivors Say?

O! that this too too solid flesh would melt,
Thaw and resolve itself into a dew.
Shakespeare

What does it take to become a hero? We hear inspiring stories of people who rush into a flaming building to save a child, or jump into a raging torrent to save an elderly person from drowning. When one man, woman, child—even a furry four-legged creature like Lassie—risks life and limb in an act of unselfish rescue, "heroic" is an apt term for it.

On a personal level, a patient who believes that a doctor, midwife, paramedic, etc. has rendered lifesaving professional service feels grateful. If the caregiver was also kind, compassionate, and available above and beyond the call of duty, the patient may later declare to friends and family, "That doc is my hero!" or "I owe her my life."

In these pages you will meet a few patients whose prostate cancer I treated. To me each of them is a hero, a trailblazer marking the way for a new treatment that eased their concerns about quality of life as well as resolving the disease that threatened their bodies.

Allan Stam, M.D.

Mentors are richly rewarded when their protégés fulfill their potential. The only time I doubted Allan's faith in my abilities occurred when I had trouble getting into medical school at the height of the medical school crunch in the mid 1970's. Discussing my latest rejection notice Allan looked at me and said, "Have you ever thought about becoming a dentist?" As it would turn out, luckily for both of us, I did finally get into medical school.

Dr. Stam is a former research professor at Harvard and past director of a Boston emergency room for three decades. His story is unusual because he was my mentor long before he became my patient! He first met me when I was a young collegiate. I had joined the Harvard whitewater kayaking club that Dr. Stam was the faculty advisor for and I did become an accomplished kayaker but as Stam tells it he noticed other talents I was yet to become aware of. As he described it I was "very bright and vaguely interested in medicine. I took him under my wing and mentored him." We became close friends as Allan set the ground rules for how a career in medicine needed to be approached. To him, a medical career was like running a marathon: when you were finished you should have used all your energy because a physician's responsibilities are too great to leave anything on the table. I was proud as I advanced through my medical career knowing that Allan also took pride in my accomplishments. He followed my research in cancer treatments.

The first time that a procedure I developed would impact one of his family members occurred early, just after finishing my fellowship. His daughter's mother-in-law was diagnosed with metastatic liver cancer from colon carcinoma. Her cancer was unresectable and she was given approximately six months to live. Allan called me (I was in Pittsburgh at Allegheny General Hospital, the Medical College of Pennsylvania at the time). After reviewing her x-rays I thought we might be able to help, but it would have to be quick since her tumor burden was just at the margin of what I thought we could do without her hurting her. Allan was present in the operating room when we explored her abdomen and unfortunately found that her disease was too extensive to destroy using cryosurgery. That was when Allan and I came up with a unique idea. Perhaps if we placed a catheter into the artery that supplied her liver to administer chemotherapy in higher doses than possible intravenously, we could shrink the disease enough to operate successfully later. This approach had been used to palliate patients (relieve their symptoms without curing them, as in chronic or terminal cases), but never had been used as far as we knew to downsize tumors to make them surgically treatable later. After six months of chemo we were able bring her back to Pittsburgh re-operate and ablate the now-reduced tumors. This has now become a fairly standard approach for downstaging patients with liver cancer.

Our mutual patient lived seven years after that initial operation before she succumbed to cancer that recurred outside her liver. As Allan tells it, "I followed Onik's innovative treatments, including taking patients that others had given up on. I was impressed by how many times he extended life for terminally ill patients, and sometimes cured them altogether. He reached a point where people from all over the world would seek him out." Much later the unthinkable happened. Allan Stam was diagnosed with an elevated PSA indicating he might have prostate cancer.

I had him come down to Florida where I had recently moved (to paraphrase the notorious bank robber Willy Sutton, "That's where the prostates are!") for an MRI and ultrasound guided biopsy. He had a small tumor that could be seen extremely well on both the ultrasound and the MRI. It was only that area of radiographic abnormality that came back positive for cancer. With only one core positive, his own doctor urged an immediate prostatectomy. He jumped into the deep end of the research pool, and was not happy with what he found. He read all the journal articles indicating incontinence and impotence as complications of RP. When Allan asked, "What should I do?" it didn't take me long to explain to him my concepts for a male lumpectomy using cryosurgery. He was a perfect candidate for this type of treatment since his tumor was so small and well seen that it could easily be targeted directly. Since Allan was such an important figure in my life, this situation tested my commitment to the concept of male lumpectomy. Could I honestly look him in the eye and tell him that if I were in his shoes this is the treatment I would be having? *No problem!*

Dr. Stam flew to Florida for the procedure in April 2001. He was in and out the same day, did not have to wear a catheter, and was both potent and continent immediately after the procedure. His PSA is stable and he considers himself cancer-free. Looking back he reflects, "I was very fortunate to have access to Gary. He's a very innovative guy. Someone I mentored developed a technique that was used on me. It was ironic that I trained him and it all came home to roost."

Larry Junker

Larry Junker is of the opinion that prostate cancer has reached epidemic proportions. At least 22 of his friends have had to deal with it, and he is supportive as they seek solutions. High school friends characterized him as a bridge builder. Little did they know that he would build bridges to the stars! Junker became a facilities engineer for NASA; now retired, he has embarked on his own growing business enterprise.

His normally low PSA of 1.5 was never a cause for concern, but during a routine DRE his doctor detected a small mass. An ultrasound revealed nothing, and Junker was understandably reluctant to undergo a biopsy. However, he agreed to a 10-needle biopsy. What a shock when the report showed one positive sample in the left peripheral zone! Though small, he was at high risk for cancer recurrence: Gleason grade 4 + 3, stage T2b.

After the initial trauma, Junker began an intense study to understand the disease and his options. He brought his engineering expertise to bear on the analysis. Like many patients who conduct meticulous research, Junker became a lay expert. He understood the technological and biological effects of accurate, extreme cold in killing cancer. After much review and many discussions, he decided on cryotherapy. An internet search led him to me, and we had our first consultation about two months after the diagnosis. Junker brought his biopsy report, bone scan results, and a list of questions. His wife and a trusted friend accompanied him. The meeting lasted over two hours. We discussed the procedure, the equipment, Junker's physical condition (irritable bowel syndrome and mild hemorrhoids) and the recovery process. By the end of the consultation his decision was firm: "Cryosurgery is the way to go." The difficult part was over and a load was lifted from his shoulders.

I then did a more definitive 15-needle biopsy to explore the regions of the nerve bundles and seminal vesicles. It confirmed the earlier findings, and the tumor appeared confined to left side. After reviewing all factors, Larry and I concurred that a partial freeze could be performed and thereby spare one of the nerve bundles to preserve sexual function. The procedure was a success.

Following his recovery, Junker underwent urinary function exams and PSA tests every three months, and a biopsy at twelve months revealed no cancer. What a year in a man's life! Since then, Junker's PSA has been monitored regularly. Another biopsy at two years again revealed that Junker was cancer-free. His PSA has been stable at 0.5 and semi-annual testing keeps his mind at ease. He has additional confidence knowing that virtually all treatment options (RP, radiation, repeat cryo) are still open should cancer recur. He sums up his satisfaction with focal freeze by declaring, "I am strong and able, continent and potent." What more could a survivor ask?

Len Capp

Len Capp says he's a bit of a Pollyanna. Having had a year of medical school, he has a healthy skepticism about rushing into treatment, and a quest to know all options. His diagnosis of prostate cancer seemed to come out of nowhere when he was in his late 50's. With a very small and moderately aggressive tumor, he got mixed messages as he traveled from his New Jersey home to various states in search of expert diagnosis and treatment advice. "Things weren't exactly fitting tongue-in-groove, which told me that prostate cancer treatment was not that clear. For things as far apart as watchful waiting and surgery to both be considered legitimate in my case, there shouldn't be that big a hole between those two options." Subsequent opinions shed no more light on his cancer—one pathologist wasn't sure he even had it from the start. Capp was suspicious of both surgery and radiation; no one had mentioned cryotherapy as an option. He decided not to rush into a decision, and committed to becoming as healthy as possible.

He embarked on a program of dietary changes, especially avoiding fats. He took antioxidants, stopped caffeine, and began regular exercise. He felt great! His PSA stabilized at 3.5 for the next nine months. He began to feel in control, and still surfed the internet for any and all information. By the end of the first year, he and his wife felt it was time to revisit the treatment decision. He reviewed radical prostatectomy, including robotic laparoscopic RP, and all forms of external beam radiation and seeds. He was leaning toward seeds despite his reservations about radiation when his wife found an article about cryo on the internet. "Look at this!" she exclaimed. That began another round of research with a few doctors repeatedly mentioned. I was one of them.

Still eager for a better diagnosis, Capp learned of Dr. Fred Lee in Michigan, renowned for his accurate diagnoses using color-Doppler enhanced ultrasound. He scheduled a trip to Michigan for a consultation with Lee. He was amazed as he watched the monitor while Lee showed him the cancerous area and took a biopsy from the site. Before Capp left, Lee told him that he could offer total gland cryoablation.

Knowing that total cryo would initially leave him impotent, possibly permanently, Capp emailed me to learn if he would qualify for focal freeze. I responded that he would have to have a Prostate Cancer Mapping Biopsy. Len was now well into his second year with the disease. He scheduled a trip to Florida for the procedure. "I was knocked out. It was quick and painless." I took almost 40 cores from the 'good' side and they were all clear. I didn't bother with the cancerous side, because I was confident of Dr. Lee's results. I wanted to confirm that if I only froze the positive half, he'd be as safe as technology allowed with respect to the other side being clean. Capp returned to his home to wait for the results, still searching the internet for anything on cryo. He was struck that there was much more information available than a year earlier.

When I called with the news that he was a candidate for the Male Lumpectomy he immediately said, "Let's do it." When his cryo was scheduled, Capp was thrilled to have the company of his wife and two sons. He went in for the procedure late on a Thursday morning, and was mostly out of it for the afternoon. He had some discomfort, but oral medication alleviated it. On Friday, he left the hospital for the hotel where he and his wife would spend the weekend before the trip back to New Jersey.

His hotel experience restored a sense of normalcy: he ate nachos, read, took a walk. Looking back at the whole experience, he feels like he had the least invasive treatment available. "I had six stitches. That was the whole thing." When he hears former cancer patients talk about being survivors, it's not a concept he relates to. "The idea of survivor is strange. We have a culture of survivors and support groups. I'm not a survivor. Other people are survivors. I'm done with prostate cancer. It's over."

Men's voices

People continue looking for solutions, and then they call Onik. (AS, New Hampshire)

It was concluded that the malignancy most probably had not escaped the prostate. After discussion of the overall situation it was determined that a partial freeze procedure could be performed and thereby spare one of the nerve bundles. That meant that I would not lose sexual function. I quickly opted for that. It worked. Dr.Onik is watching my case very closely. (LJ, Florida)

Having had a focal freeze, I feel like I'm a lucky son-of-a-gun. I got the least invasive treatment by a long shot. Least side effects, least rehabilitation, and the results are as good as any other. (LC, New Jersey)

Summing It Up: Clothes Make The Man

CTDA material reprinted with permission of The Custom Tailors & Designers Assn., www.ctda.com.

Have you ever had a suit or shirt made just for you? The Custom Tailors and Designers Associates (CTDA) is all about the art of customized clothing. If their principles were translated into the language of cryotherapy, how would they read?

CTDA: There are times in a man's life when he cannot afford to feel second best.

Cryo: There are times in a man's life when he cannot afford to feel second best.

CTDA: Simply stated, custom tailoring is selection, not compromise – fit, not finagle. It will be styled and fitted properly – by choice, not by chance.

Cryo: Simply stated, custom freezing is selectively focused, without compromise – targeted, not finagling with pelvic anatomy. It will be shaped and sized properly – by an experienced doctor assisted by computerized algorithms, not by guesswork.

CTDA: Always, selection and fit are the key advantages. But, more importantly, they will fit the man's personal preference – giving him full comfort and confidence in his appearance.

Cryo: Always, immediate cancer death and a margin of safety are the key advantages. But, more importantly, they will fit the man's personal lifestyle – giving him full comfort and confidence in his masculinity.

CTDA: It's a good idea to narrow the field before approaching your tailoring consultant. Look over the styles. Try to decide which might suit your tastes and your life goals best. Be honest. It's important to have some ideas about what you need beforehand.

Cryo: It's a good idea to narrow the field before making an appointment with a cryosurgeon. Learn about all treatment options. Try to decide which might suit your needs to be cancer-free and preserve your lifestyle best. Be honest about your concerns. It's important to have some ideas about what you need beforehand.

CTDA: Together, you and your consultant will select the styling options that will promote appearance, lifestyle, and comfort. Personal styling can enhance your best physical qualities and soften any shortcomings. No compromises. No major alterations to reshape or refit an "average" size off the rack. In every detail, it has been truly created for one man.

Cryo: Together, you and your physician partner will select the freezing options that will promote safety, lifestyle and comfortable recovery. Personal cryo can enhance your best physical function and soften any concerns. No compromises. No major surgery to recover from or radiation with its "average" scatter effect. In every detail, the iceball has been truly created for one man.

CTDA: Once a man experiences custom tailoring, he is never satisfied with the second best again. Once he has been comfortably fitted in his first custom suit, he will never worry about tricky alterations or refitting again. Custom tailoring is clothing of choice, not chance.

Cryo: Once a man has a customized focal freeze, he is satisfied that he has second options if necessary. Once he has been successfully treated in his primary cryo, he will never worry about tricky alterations to his internal anatomy or refitting his lifestyle. Custom freezing is a treatment of choice, not total-gland chance of side effects.

CTDA: Yes, once a man experiences custom tailoring, he'll understand what successful men around the world have known for centuries; it's so personal and so convenient, why did I wait so long?

Cryo: Yes, once a man experiences tailored cryo, he'll understand what happy former patients around the country have known for years; it's so personal and so controlled, why would I wait any longer?

Chapter 7
Achieving The Vision Together

All great discoveries are products as much of doubt as of certainty, and the two in opposition clear the air for marvelous accidents.
M. Helprin

One fine day a Postal Service carrier walks into the office suite where Dr. Goodbuddy and his urology partners are seeing patients. The receptionist smiles a hello as she takes the bundle of mail from him. She dumps the junkmail into the recycling bin, sorts through the keepers, and stuffs Dr. Goodbuddy's packet into his mailbox.

That night, in the halo from Dr. G.'s bedside lamp, his wife fondly kisses her partner of 29 years. As she plumps her pillow, she peers at the medical journal he's holding. "Working late?" she teases. "Where's that pulp novel you couldn't put down?" "Mm," he grunts. She settles back, grateful that he finally has a regular work schedule. As she drifts off into serene sleep, she anticipates a good family time when she and Dr. G. take the coming weekend to visit their son at his medical school.

Dr. Goodbuddy sees no reason at this point to confess that he's a little restless because his annual physical revealed a rising PSA. In one year it went from 2.3 to 3.1, not overly alarming in itself. He prides himself on practicing what he preaches. His DRE and ultrasound were negative, so he's planning to do what he advises men at the hospital's free prostate cancer seminars. "Never panic. Routine PSA tests can fluctuate naturally over time. Cancer isn't the only factor that can raise PSA levels. Wait about six weeks and get another test. If it hasn't gone down, or has risen slightly"—

here he always smiles reassuringly—"call my receptionist to book an appointment and we'll look into it." Still, as his wife snuggles against him, he reflects on the dark irony that he may now have something in common with many of his patients: prostate cancer. When and how would he tell her?

In order to steer his mind away from these concerns, he focuses on the journal's table of contents. A title leaps off the page: "The Male Lumpectomy: A Rationale For Using Focal Cryoablation To Treat Prostate Cancer." He doesn't recognize the author's name. He flips to the article. As he swiftly reads, his reactions are mixed: anxiety, doubt, curiosity, common sense and other feelings eddy around a core of inner hope. The data looks exciting. Now he has a different reason for not sleeping. He can't wait for morning to pursue more information about what he has just read. He ponders whether to bring his laptop on the weekend's excursion. His wife won't be happy about it, so he'll have to tell her about his health concerns, and maybe tell their son too. However, this article gives a basis for reassuring them. For reasons both personal and professional, he marvels that this issue arrived today. Coincidence? Fate? Maybe cryotherapy is something he should learn to do so he can offer it to his patients. He feels grateful now too and caresses his wife's hair. She stirs and spoons against him. His spirits lift as his mind opens to new dreams.

This scenario is fiction but it repeats itself all the time in real life. We have to ask ourselves, "Why would a disproportionate number of patients for this new approach to treating prostate cancer be physicians?"

Desire, dreams and science

In 1776 a group of disgruntled men declared their colonies independent of the British Empire. That same year the English author Samuel Johnson wrote, "Life is a progress from want to want, not from enjoyment to enjoyment." Desire springs from want, and in turn creation springs from desire. We are spurred by longing. If one person's need plants the seed for a beneficial outcome, many who shared the want but not the wherewithal will profit."

Medicine itself progresses this way. As individuals and as a society, we want to be free from cancer. As you have seen, the story of the Male Lumpectomy is a wonderful example of forging a creative technological alliance into a unique way of satisfying this want. The voyage has not been smooth. Along the way, we have met disappointments, obstacles, even betrayals that discouraged us. At times we paid a personal price in our quest for a middle ground in prostate cancer treatment and our patients paid a price while we developed "good judgment" but great inner desire is essential to fuel a long drive toward achievement.

New dreams constructively challenge the medical establishment, which is rightly conservative toward promising breakthroughs. A dynamic seesaw between appropriate caution and the edginess of the dream marks a testing period. As the outcome hangs in the balance, the dreamer looks beyond his own brainchild for support from similar precedents. We have seen how breast lumpectomy altered the treatment landscape for women with breast cancer (BCa). Each doctor who excised a lump before the efficacy was proven in huge clinical trials took a chance with a woman and those who loved her. Every woman who gave informed consent to use her body as a living laboratory ventured into unknown territory. Along the way, many were made to feel guilty for not choosing the "gold standard," but their determination and desire were ultimately rewarded. Breast lumpectomy was vindicated because of women who insisted: "I want my breasts." If a person wants something badly enough, he or she embraces creative risk to achieve it. When enough people want the same thing badly enough, collective desire impels them toward a critical mass that changes the status quo.

We doctors too have creative longing. As we listen to patients, read journals, attend conferences, and rack up continuing education hours, we move from need to desire to creation. Doctors *want* the best for our patients. Men *want* to know about partial gland treatment. As they learn, they *want* to choose it, because they *want* quality of life as well as life itself. "Life is a progress from want to want..." Male Lumpectomy is a result of this. When enough men *want* focal freeze and enough doctors respond, it will produce evidence that affects how prostate cancer is understood and treated.

Today's revolutionary patients

Progress in breast lumpectomy inspired me to develop an equivalent prostate procedure. BCa organizations provided an enviable political model to their male counterparts. Prostate cancer organizations have tried to find a formula for equal success in fundraising. They mobilized to capitalize on the brilliant idea of a small piece of paper, as described on the Men's Health Network website:

The breast cancer stamp passed Congress and was signed into law in the fall of 1997. The stamp went on sale in mid-1998 and has been a great success with over

$8 million raised for research in the first year of sales. Each sheet of 20 stamps includes the National Cancer Institute's Cancer Information Service number (1-800-4 Cancer) as the call to action... Introduction of the breast cancer stamp was followed immediately by stamp events in hundreds of post offices.... Almost all of the 40,000 post offices promoted the campaign through posters in their lobbies, distribution of free breast cancer brochures, radio and television public service announcements and other community outreach activities.

Unfortunately, the PCa stamp was short-lived for several reasons: Congress passed the bill but handed off responsibility for issuance; other medical causes competed with PCa; the American Cancer Society did not support the stamp's message about annual PSA testing, the value of which is considered debatable. In contrast, women have become even better at mobilizing political will. Their efforts often begin with informal but intimate conversations. From carpools to cocktail parties, women seem comfortable sharing their physical concerns with each other. An example might go like this:

Jen: Wow, Barb, your skin looks great! Are you using something new?

Barb: Not exactly. I've started peri-menopause, and the thought of aging is totally depressing. So I decided to get weekly facials to be nice to myself. I see a difference, but what about the rest of me? I swear I'm totally drying up.

Jen: Tell me about it! I went off hormones when the news came out about that study. I don't know how my family puts up with me. Hot flashes, rage attacks—I can hardly stand myself. I look in the mirror at these crows' feet, my libido is zilch, and now Bill discovered Viagra so I have no excuse. Where's the justice? You know a good lubricant by any chance?

Barb: Liz and I were just talking about the same thing. How are we going to make it through menopause without ending up in divorce court? We should start a wine-and-cheese group for our sanity. With low-fat cheese, of course.

Jen: Yeah! I'll call Alyssa and Marie, you call Liz and whoever else. How about my house this Sunday afternoon, say three-ish?

It will be no surprise if they launch an estrogen research campaign fund!

We guys, on the other hand, are more likely to chat about a body of ideas than bodily functions. When we're not talking about sports, we might tackle things like economics, politics, or science. Organized political action, possibly spearheaded by a bold leader, may arise from spirits that take a while to ferment and distill. Take two famous examples. In 1517 when Martin Luther's "95 Theses" catalyzed religious reform, his document embodied a wide current of theological criticism, not just his own. In 1776 the Declaration of Independence galvanized a rebellion. Thomas Jefferson was the main author, but he was the spokesperson for shared new ideas of human rights and equality. It was as if Luther and Jefferson each uncorked a collective bottle containing a powerful genie who could grant a wish: **diversity** of belief in one era, political **liberty** in the other.

A new wave of awareness about PCa is brewing among men who have it and men at risk of developing it. Guys may not kibitz about their bodies over wine and cheese, but they have a multitude of ways to seek and share information: news, books, the world-wide web, email groups, on-line chat rooms and bulletin boards, organizations, even direct-to-consumer advertising via print, radio and TV media. And of course, their women bring news home—and sometimes "nag" about that annual physical! Despite government anti-screening policy, PCa groups encourage annual PSA testing because the earlier PCa is diagnosed, the more options a man has. As this informational trend grows there are new sparks of healthy, rational discontent. Demand for change is brewing.

The number of women who demanded less mutilating breast surgery became so large that the medical community responded with data on unifocal BCa and breast lumpectomy backed up by a safety net. We see the same trend with men: increasing education and empowerment, asking for detailed diagnoses, and giving informed consent to innovative but not yet widely tested treatments. As the ranks swell, it fires my imagination, and other doctors like myself, to articulate the increasing patient need for **diversity** of treatment options and **liberty** from cancer as well as the side effects of treating it.

I collaborate with my patients in generating progress. As partners, we trust the growing evidence of unifocal PCa. We trust the technology to deliver a lethal yet targeted blow, plus an added safety margin. We trust that by tailoring the freeze to a percentage of the gland, male dignity and undiminished quality of life are preserved. As doctor and patient team up in the march of progress, perhaps future men will look back and view the Male Lumpectomy as a contribution to what they well might call a "Prostate Reformation."

Men's Voices

It was not easy to look into cryo at the beginning. For the first five months I only spoke to people who were doing RP and radiation. We're all so affected by the media. There was just so little on cryo. I had to focus and not let the other things I heard about cryo affect what I was reading. Within a year, I was struck by what seemed like an explosion of information on cryo. My wife was absolutely supportive, especially because of what she had been through. We were both really sensitized to being our own advocates, having a decent brain, looking through things, not being afraid to ask questions. (LC, New Jersey)

I did cryosurgery guided by Spectroscopic MRI for about 7 years. I also have done limited cryosurgery to preserve the nerves for many years. This is not news to me. Instead of cryosurgeons admiring and congratulating themselves that cryosurgery is approved at last as a treatment for prostate cancer disease, we also need to make progress and protect the patient by maximizing the protection of quality of life issues. We should focus on treating the patient and not just on treating his prostate. (I. Barken, MD)

It's really in the men's interests and the interests of the medical community to stop cryo from being pigeonholed. It's the flexibility of the technology that allows patient-specific treatments that is not available in other treatments. (D. Rukstalis, MD)

Summing It Up:

Each person—professional, patient, survivor or loved one—who has an intimate encounter with cancer enters a classroom with a strict teacher. There are many lessons, lots of discussions (participation counts toward your grade!), room for error, and no long boring lectures. The class is experiential, so participants learn by doing. There's no such thing as auditing. The course is only available for Life Credit, but there are no grades, and no right or wrong answers on tests like Attitude, Responsibility, Nurturing and Empowerment.

This book is not the course textbook. It is the story of how a new treatment approach for prostate cancer, Male Lumpectomy, is part of an evolving kit to help students who are in the class. If it's not right for you, hopefully it will help you refine your search for the right tool.

Perhaps the most important point of this book is simply: Believe that you are making a contribution to humankind. We can all learn from you. Tell others honestly what you think and how you feel. Ask questions. Lean on others for resources. You are a Partner in Progress.

Appendix I: Additional Resources

Please note: These are pointers to Internet-based PCa resources, and should not be regarded as any sort of recommendation or endorsement. Carefully weigh the information you find, and discuss it with your own physician.

PROSTATE CANCER MAILING LISTS

NewDx
Information and support to those newly diagnosed with prostate cancer. Experienced PCa patients who want to help the newly diagnosed are encouraged to join!
 www.prostatepointers.org/mailman/listinfo/newdx

The Circle
Support for wives, families, friends, and significant others of men with prostate cancer -- and, of course, the men themselves.
www.prostatepointers.org/mailman/listinfo/circle

IceBalls
Information and support to those interested in cryosurgery for prostate cancer.
www.prostatepointers.org/mailman/listinfo/iceballs

Prostate Cancer and Intimacy
Frank and open discussion of the sexual issues surrounding PCa.
www.prostatepointers.org/mailman/listinfo/pcai

Spirit
For those who want to share spiritual support as they live with PCa.
www.prostatepointers.org/mailman/listinfo/spirit

Prostate Cancer Action Network
A forum for discussing important issues affecting the care and treatment of PCa survivors, and ways to bring about needed change.
www.prostatepointers.org/mailman/listinfo/pcan

Prostate Cancer and Gay Men
http://health.groups.yahoo.com/group/prostatecancerandgaymen/

PROSTATE CANCER WEB SITES

CryocarePCA (Prostate Cancer Advocates)
http://www.cryocarepca.org

Prostate Pointers
http://www.prostatepointers.org

Information for the newly diagnosed
http://www.prostatepointers.org/newlydx.html

Us TOO
http://www.ustoo.org

Prostate Cancer Profiler
http://www.cancerprofiler.nexcura.com/Interface.asp?CB=66

Prostate Cancer Research Institute (PCRI)
http://www.prostate-cancer.org/

Prostate cancer acronyms and abbreviations:
http://www.prostatepointers.org/prostate/ed-pip/acronyms.html

Prostate cancer glossary of terms:
www.prostatepointers.org/prostate/ed-pip/glossary.html

Information provided by Charles Myers, MD
www.prostatepointers.org/cmyers

Information provided by Israel Barken, MD
www.prostatepointers.org/barken

Physician to Patient
www.prostatepointers.org/p2p

IceBalls
www.prostatepointers.org/iceballs/

The PCAI list of ED specialists
www.prostatepointers.org/pcai/ed.html

Organize your PC digest:
www.prostatepointers.org/p2p/pcd.html

Prostate Cancer Stages
www.phoenix5.org/staging.html

National Prostate Cancer Coalition
www.pcacoalition.org/

PAACT
http://www.paactusa.org/

Phoenix5
http://www.phoenix5.org/

Malecare: lecture transcripts, English and Spanish NYC support
groups. Also sensitive to gay men's issues.
http://www.malecare.com

Canadian Prostate Cancer Network
http://www.cpcn.org/

BOOKS

American Cancer Society's Complete Guide To Prostate
Cancer, Ed. By D. Bostwick, D. Crawford, C. Higano, and M.
Roach III, 2005

Centeno & Onik, Prostate Cancer: A Patient's Guide to
Treatment, Omaha NE: Addicus Books, 2004

Hennenfant, Bradley, MD, <u>Surviving Prostate Cancer Without Surgery</u>, Roseville Books, 2005

Strum & Pogliano, <u>A Primer On Prostate Cancer: The Empowered Patient's Guide</u>, Hollywood FLA: The Life Extension Foundation, 2002.

RECOMMENDATIONS

Spread the word: a genetic connection has been established between prostate cancer and breast cancer. If your mother, aunt or sister had breast cancer, you may be at greater risk for prostate cancer. If you know a woman whose father, uncle or brother had prostate cancer, she may be at greater risk for breast cancer.

Include your significant other in all treatment decisions. Discuss both of your feelings honestly, especially about possible side effects of treatment such as incontinence and impotence.

Bring a loved one and a notebook or tape recorder with you to all medical appointments.

Request copies of your medical records, test results, pathology reports etc. You are legally entitled to access to all files, and to have copies for your personal files.

Appendix II: The Prostate Cryotherapy Procedure

Cryotherapy (also called cryoablation, cryosurgery, or cryo) as a treatment for localized prostate cancer is gaining favor among patients and doctors for primary (first) treatment as well as salvage treatment if cancer recurs after radiation therapy. It uses lethally cold freezing to destroy prostate tumors. This quickly kills cancer without the risks of major surgery or radioactivity. Recent advances in equipment and methods protect surrounding healthy structures during the procedure.

Remember that all medical procedures involve some risk, and each patient is unique. Your age, medical history, stage and location of cancer, attitude, family support—these are factors that can affect your treatment experience. With cryo, as with all treatments, be sure to have a thorough discussion with your doctor. Never hesitate to ask your physician to explain or clarify what to expect before, during and after the procedure. The following is a general description of prostate cryotherapy. It is not intended as medical advice, and is not a substitute for the specific way each doctor and hospital administers the procedure.

Before the scheduled cryotherapy procedure, a doctor will describe the preparation steps you must do at home, and what to have on hand during the recovery period, such as icepacks. He will indicate whether your cryo will be performed on an outpatient or inpatient basis.

Total Gland Cryo

While the patient is anesthetized (local or general) the doctor uses transrectal ultrasound guidance to insert 6-8 slender cryoprobes into accurate locations in the prostate gland. The probes are further guided and stabilized by a stationary grid positioned directly in front of the perineum, or area between the scrotum and anus. This is the skin area penetrated by the probes as they are inserted into the prostate.

Since the urethra passes through the prostate gland, a warming catheter is inserted through the penis into the bladder. This circulates sterile heated saline solution to protect the urethral lining from freezing. Thermal sensors keep track of the temperature in and around the prostate to avoid damage to the bladder and rectum. While computerized calculations assist in placing probes to avoid rectal damage, some cryosurgeons also inject sterile saline solution into tissue (Denonvilliers' fascia) between the prostate and rectum to temporarily widen this space. The fluid is naturally absorbed and excreted after the procedure.

The length of each cryoprobe is insulated except for the tip. When all is ready, liquid (pressurized) argon gas is circulated within the probes, forming "teardrop" shaped iceballs at the tips. These iceballs overlap into one solid, prostate-shaped mass, freezing the prostate tissue to −40° C. A safety margin of ice extends beyond the gland into the tissue surrounding it, or prostate bed. This is done as extra insurance against the spread of prostate cancer into the body. The doctor clearly "sees" the iceball on the ultrasound monitor, so he knows that he has frozen exactly what he intended. Additionally, the thermal sensors indicate the temperature in and around the prostate confirming that the cancer cells have been destroyed.

Once the freeze is complete helium gas replaces the argon to thaw the tissue. The freeze/thaw process is repeated once more. The cancer and its blood supply are now destroyed as the cells both rupture and dehydrate during the process. The dead tissue is re-absorbed over time or remains in the body as scar tissue and poses no other health threat. The procedure takes about two hours; patients either go home the same day or spend a night in the hospital.

While many patients resume normal activity in less than a week, some patients may experience temporary bruising and swelling. A urinary catheter is left in place for 1-3 weeks of internal healing, then removed.

Less than 1% of patients report incontinence following cryosurgery. Since the nerve bundles are adjacent to the prostate gland, most patients will be impotent if the entire prostate is frozen. However, 47% of previously fully potent men regain erectile function, usually in 12-36 months. Regardless of erectile ability, all patients are capable of orgasm following cryo, as with all prostate treatments, since a different nervous system controls sexual climax.

Cryoablation has other advantages as well. Recovery is rapid. Most patients return to their normal lifestyle quickly. Unlike radiation, cryo can be repeated if prostate-confined cancer recurs (comes back). Furthermore, it is the *only* Medicare approved treatment if prostate-confined cancer recurs *after* radiation therapy (external beam or brachytherapy).

Long-term statistics show that cryosurgery is at least as effective as radical prostatectomy and radiation for low-risk cancers, and has better success rates than surgery and radiation for moderate to high-risk cancers. Cryo is more effective than external radiation or seeds if the cancer is confined to the prostate gland (up to stage T3 tumors.)

Cryoablation for prostate cancer is Medicare approved for both first-time occurrence and post-radiation recurrence.

Partial Prostate Cryo And Male Lumpectomy

The technique for freezing any portion of the prostate less than 100% involves using fewer probes. The planning of how many probes and where to insert them depends on the size, location and risk factors of the tumor. In other words, the cryosurgeon can "sculpt" as much ice as necessary to freeze the tumor plus a safety margin.

In all other respects, the procedure is virtually identical to total gland cryo. In the case of a focal freeze such as Male Lumpectomy, the use of a catheter after the procedure may not be indicated, or may be only a few days.

Patients must be carefully qualified by their physician for this approach to ensure that no clinically significant prostate cancer is detected outside the area to be frozen.

Nerve-Preserving Or "Nerve Warming" Cryo

Some cryosurgeons use an approach that freezes all or most of the gland but spares one or both neurovascular bundles from the most aggressive ice. The procedure is similar to total gland cryo. In this case, however, a cryoprobe that circulates helium is placed adjacent to one or both bundles, along with temperature monitors. The intent is to avoid a hard freeze of the nerves to preserve erectile function. Some patients experience an immediate return of potency, others find that the nerves regenerate over time. Preliminary studies show 55% potency at six months follow-up.

Hundreds of doctors in the U.S. offer cryoablation. For more information or a doctor referral call toll-free 1-877-PCA-CRYO (877-722-2796). For patient experiences, doctor profiles and journal articles visit www.cryocarepca. org.

Appendix III: Indications for Prostate Cryoablation

Cryoablation (freezing) as a treatment option for prostate cancer is appropriate under the following conditions, with a physician's consultation:

Primary (no previous) whole-gland local treatment
- Stage T1 – T3 disease
- A thorough and appropriate diagnostic work-up has determined that prostate cancer is localized (no remote metastasis to lymph nodes, bone, or other organs)
- Age, other medical conditions, or previous medical/radiation treatments rule out radical prostatectomy or primary radiation
- Cryo is statistically more effective against high risk tumors that may not be amenable to prostatectomy or radiation, i.e. PSA greater than 9, Gleason grade equal to or greater than 7, Stage T2b or greater
- Patient understands possible side effects and recovery process associated with total gland cryoablation

Salvage treatment for recurrent prostate cancer following previous radiation or cryo
- Diagnostic tests have determined there is no remote metastasis
- Only salvage therapy approved by Medicare for radiation recurrent disease
- Patient is concerned about urinary/rectal side effects of salvage prostatectomy or salvage radiation
- Patient understands possible side effects and recovery process associated with total gland cryoablation

<u>Less than 100% gland cryo with intent to preserve potency</u>
- Biology of patient's disease has been diagnosed with a high degree of confidence
- Physician is qualified to offer potency-preserving cryo
- Physician has acquainted patient with attendant risks and safety measures
- Patient is motivated to observe physician's follow-up protocol scrupulously

Disclaimer: Statements contained in this book are not medical advice, nor intended to substitute for same. As with all questions of a personal medical nature about prostate cancer, treatment for this disease, or any other medical question consult your doctor.

Dr. Gary Onik can be contacted at his office in Celebration FL at (407)303-4228 or online at <u>onikcryo@aol.com</u>.

Acknowledgments

We would like to acknowledge the administration of Florida Hospital/Celebration Health who have actively supported Dr. Onik's prostate cancer research program; Drs. Vaughan, Brady, Dineen, Brunelle, Narayan, Barzell and Lotenfoe, the urologists that work with Dr. Onik and his coauthors on the research papers dealing with "The Male Lumpectomy;" all those whose valuable expertise, experience and wisdom are quoted in these pages; Endocare, Inc. (Irvine, CA) for permission to use their illustrations; finally, as always Dr. Onik particularly acknowledges the contribution of his long term collaborator and friend Boris Rubinsky PhD whose brilliant engineering contributions kept image guided tumor ablation advancing despite the many problems encountered.

Gary Onik M.D. and Karen Barrie M.S.

LaVergne, TN USA
10 September 2009
157468LV00003B/281/P